Whose View of Life?

Jane Maienschein

Whose View of Life?

EMBRYOS

CLONING

AND

STEM CELLS

HARVARD UNIVERSITY PRESS

CAMBRIDGE, MASSACHUSETTS

LONDON, ENGLAND

2003

Library of Congress Cataloging-in-Publication Data

Maienschein, Jane.
Whose view of life? : embryos, cloning, and stem cells / Jane Maienschein.
p. cm.
Includes bibliographical references and index.
ISBN 0-674-01170-8
1. Human embryo—Research—Political aspects—United States.
2. Stem cells—Research—Political aspects—United States.
3. Human cloning—Research—Political aspects—United States.
I. Title.

QP277.M356 2003
176—dc21   2003051084

With a million thanks to Rick and Marie

Contents

Figures

Acknowledgments

This book has its origins in opportunities that came to me from two directions. First, a series of National Science Foundation grants allowed me to explore the history and philosophy of organismal biology and embryology. At the same time, Arizona State University's then-President Lattie Coor, Provost Milt Glick, Dean Gary Krahenbuhl, and Biology Department Chair James Collins created the opportunity for me to serve as science advisor for Arizona Congressman Matt Salmon. That position took me back and forth between Washington and Arizona for two years (during the 105th Congress) and opened the door to wonderful experiences and insights. Then a dozen undergraduates came along, and together we explored issues of scientific literacy and finally produced an editorial in *Science* in 1998 ("Scientific Literacy," no. 281, p. 917) as well as a longer paper in *Science Communication* ("Commentary: To the Future; Arguments for Scientific Literacy," vol. 21, pp. 75–87). The latter has served as the definition of "scientific literacy" for the National Science Foundation's *Science and Engineering Indicators*. I wish to thank Suzanne Reuss, Patrick Burkhardt, Robert

Barnhill, and Jon Fink for providing important support during this adventure in advising.

During that time I also served on the Program Committee at the American Association for the Advancement of Science, guided by Michael Strauss, and as chair of the Advisory Committee for the National Science Foundation's Directorate for Social, Behavioral, and Economic Sciences, appointed by Cora Marrett. Given my involvement in these varied endeavors, I was able to view at first hand the debates over important issues such as whether to regulate or fund cloning and stem cell research; it was an amazing opportunity to pull multiple perspectives together. Michael Fisher at Harvard University Press encouraged me to record my experiences and to think about ways to turn them into a book; he has been a wonderful friend and guide along the way.

Since then, I have participated in many educational programs for the community through the outreach programs of the president's office at Arizona State University. Elva Coor and her advisory committee persuaded me to give a science course for community supporters, even though others were convinced that the topic of cloning would prove too "heavy" for the program. In fact, we had an enthusiastic group of adult learners, and I surely learned more than they did from the four weeks of intense discussions on the science and policy of human reproductive research and regenerative medicine.

The Federal Judicial Center led me to Judge Robert Henry of Oklahoma and Judge Helen Berrigan in New Orleans, both lifelong learners and eager supporters of educational programs for all. They each have introduced me to legal and other considerations concerning effective educational programs geared to scientists and judges together. With the Medical Humanities Program at Mayo-Scottsdale, directed by Dr. Mark Edwin, Arizona

ACKNOWLEDGMENTS

State is beginning to develop community outreach programs and opportunities for our students in bioethics and social issues.

Pulitzer Prize–winner and good friend Jonathan Weiner, along with his talented wife Deborah Heiligman, came to share their writing and their family with a select group of undergraduates in a "Science and Literature" course administered through our Honors College. They inspired us all, especially me, to get busy and write. The program was part of our growing Biology and Society Program at Arizona State, which attracts absolutely the best students in the universe, and we clearly have the most fun.

All of these activities—as well as my various professional responsibilities on governing boards for the History of Science Society, Philosophy of Science Association, International Society for the History, Philosophy, and Social Studies of Biology, and the Association for Women in Science—share common themes. Their convergence encouraged me to take a multidisciplinary perspective on issues of historical and current interest. Working in different arenas also allowed me to combine academic and scholarly commitments inside the university walls with the world of politics and policy. This is an exciting time in the history of the biosciences, and I thank everyone who has helped me to be part of it all.

The book came together when I took a leave from Arizona State thanks to Dean David Young and undertook a senior fellowship at the amazing Dibner Institute for the History of Science and Technology at MIT. The staff and other fellows make this a most wonderful place to work. With offices looking out over the Charles River toward Boston, the fellows enjoy afternoon tea and cookies and an atmosphere of peace and quiet broken only by lively discussions and weekly seminars. My friend and colleague Garland Allen sat in the office next door, and my husband and colleague Richard Creath in the office beyond that.

We wandered in and out of each other's offices to organize lunches, to ask questions and exchange ideas, and to spur each other on in friendly competition. It is difficult to imagine a better setting for writing a book in six months.

A team of readers, including Marie Glitz, Garland Allen, Françoise Baylis, John Bonner, Esther Ellsworth, Manfred Laublichler, Mia McNulty, Matthew Shindell, Kori Wallace, and my father, Fred Maienschein, read the entire book and made absolutely indispensable suggestions. They helped me to think through the tough parts and pushed me to make the transitions and conclusions clearer. Gar and Marie read much of the manuscript more than once, and Rick Creath read through some parts in a million drafts. Their substantive and editorial suggestions and corrections have proven invaluable. My mom, Joyce Maienschein, James Collins, Julie Story, and Jim Hathaway also offered insights from their reading along the way. We have diverse views about how the human embryo research debates should turn out, and what is ethically preferable, and that is just what we should expect. It has been valuable for all to engage in discussion in an attempt to forge just the sort of cooperative effort to make effective policy for which this book calls. I have benefited from conversations with more generous colleagues and students than I can even remember.

Sara Davis and Susan Abel at Harvard University Press have both been exceptionally supportive, guiding me through the world of permissions, illustrations, page proofs, indexes, and other practical matters. I always worry about copyeditors, because some are just harder to work with than others. But Harvard assigned me to Kate Schmit, who is clearly the best. She seemed to know just what I wanted to say and what tone I was trying for, and she offered gentle suggestions that were always better than my own.

At Arizona State University, Christy Hansen helped with preparing the manuscript. Dean David Young, Provost Milt Glick, and President Michael Crow have created the kind of academic environment in which it is fun to work hard on challenging projects. They are all part of this book, which truly has been a collaborative effort. That is as it should be for a multidisciplinary project that requires input from different points of view and that, to be successful, must speak to diverse audiences.

There is grandeur in this view of life . . .

—CHARLES DARWIN,
On the Origin of Species, 1859

Introduction

On August 9, 2001, United States President George W. Bush spoke to the nation from his ranch in Texas about "a complex and difficult issue, an issue that is one of the most profound of our time." "The issue of research involving stem cells derived from human embryos is," he continued, "increasingly the subject of a national debate and dinner table discussions. The issue is confronted every day in laboratories as scientists ponder the ethical ramifications of their work. It is agonized over by parents and many couples as they try to have children, or to save children already born." The president had consulted many parties, including scientific, legal, ethical, and religious experts, doctors, members of Congress and of his Cabinet, and friends. In addition, he assured us that he had "given the issue a great deal of thought, prayer and considerable reflection." What he found, not surprisingly, was "widespread disagreement."[1]

Despite the acknowledged coexistence of competing views, President Bush felt obliged to decide whether to allow tax dollars to be used for a research program that was both full of hope and fraught with disagreement. During his campaign, he had pledged to oppose human embryo research, and he was under considerable pressure to act on those promises. We will reflect later on why this president made the decision the way he did and why George W. Bush, a self-proclaimed mediocre student and no particular fan of academic research, was in the position to decide what research the National Institutes of Health (NIH) would be allowed to fund. Why did he find so many different arguments and so many would-be experts speaking to this particular issue?

This book will provide perspective on such questions, as well as explore the answers. Human embryo research is, of course, first of all about science and about research. It is about intellectual freedom and the excitement of scientific inquiry. Yet it is also about human embryos and making decisions about how to define what we will count as a life. We live in a society where we value life, and we value science. We value protecting individual freedoms, and we value medical advances. Most important, we have a range of moral views and a range of views about science. This is not a contest between the moral and the scientific, but rather a deeper set of conflicts about how to reconcile competing views and values in a pluralistic society. How, in particular, do we decide how to define when a life begins and what the appropriate boundaries and constraints on human embryo research should be?

This is a story about biology in American democratic society. Science plays a special role in a democracy, perhaps especially in a constitutional and representative democracy. By various estimates, well over half the decisions in the United States Congress involve science in some sense and in some way. When it comes to matters that involve science, should we rely on our usual democratic decision-making system of committees and experts—or do some experts ever trump others precisely because they know the science? Who decides what counts as "sound science," which experts best know or can best explain that sound science to others, and what the proper role is for science? What is true, what is good, and who decides?

Stem cell research, like cloning and other areas of human embryo research, raises problems because it seems at its core to be about defining and defending life. Some hold that a life begins at conception (when the sperm cell fertilizes the egg cell and results in a new cell with the full complement of two sets of chromosomes). For the most extreme advocates of this view—those who hold as absolute and inviolate truths that conception initiates a full life and that every life at any developmental stage is absolutely good—any disruption of that life is immoral. Further, many absolutists believe that any disruption should be illegal and that any research that harms the individual human life should be prohibited.

To move in the direction of legislating such a moral interpretation, a self-proclaimed conservative Congress in 1993 enacted a ban on federal funding of human embryo research. This restriction came as a rider on appropriations bills, since the supporters did not have the votes to pass independent legislative prohibitions on all research. As recent funding bills, have elaborated, any

> Grant, cooperative agreement and contract funds may not be used for (1) the creation of a human embryo or embryos for research purposes; or (2) research in which a human embryo or embryos are destroyed, discarded, or knowingly subjected to risk of injury or death greater than that allowed for research on fetuses in utero under 45 CFR 46.208(a)(2) and Section 498(b) of the Public

Health Service Act (42 U.S.C. 289g(b)). The term "human embryo or embryos" includes any organism, not protected as a human subject under 45 CFR 46 as of the date of the Act, including any by fertilization, parthenogenesis, cloning, or any other means from one or more human gametes or human diploid cells.[2]

Legislators sought with these bills to define the lives that would be protected from research.

While some hold the most extreme view that any embryo research is necessarily bad, an alternative view insists that though embryos are clearly alive in some basic organic sense, each embryo is not yet a life in a meaningful sense. They maintain that there are stages in early life, and that some of those stages count as a human life and deserve protection, while others do not warrant the same treatment. They see differences as the individual developmental stages progress. Those who hold the strongest extreme of this view would allow research on embryos and even in some cases on later developmental stages, though research on these later stages after implantation in the uterus would presumably require consent of the mother. For a very few, the relevant end point for allowing research might be at birth, but for most the difference arises at some earlier stage. This might be the point when the embryo has become a fetus and has acquired all its parts, or when it becomes "viable" as an entity that can (with support) sustain life outside the mother.

This position, that a life emerges only gradually, underlies the reasoning that led to the legalization of abortion. It does not follow that everyone who holds this view would advocate open research on fetuses, but it does follow that there is a range of interpretations—and reasoned interpretations—about when a life becomes a meaningful life worthy of protection and a separate

WHOSE VIEW OF LIFE?

range of reasoned interpretations about how late in the developmental process it is legitimate to carry out research on developing human organisms.

In the United States, we have also legally accepted another middle ground. Fertility clinics are allowed (and indeed encouraged by the infertile couples who hope to benefit from the resulting medical interventions) to mix sperm and eggs in glass dishes and produce fertilized eggs. They can allow the eggs to begin cell division and then store them, destroy them, or implant them in mothers. There is no federal legislative restriction on these medical procedures. Our political acceptance of this technology shows that our society has a range of reasonable views of when a life actually begins.

The problem is how to accommodate all these different and competing views. We can do that in our usual way, by allowing different individuals to pursue their preferences. Some individuals may choose to use the *in vitro* fertilization techniques of fertility medicine, some may choose to have abortions in the first trimester or into the second, some may adopt contraception to prevent fertilization and subsequent embryo formation, while still others may choose personally not to adopt any techniques that rely on human manipulation of embryos. This sounds just fine: individual free choice. Yet, for those who believe that protecting innocent human embryos is a matter, quite literally, of those embryos' life or death, certain types of research ought not remain a matter of choice. Here we reach fundamental and deep disagreements about morality and what it means to be a civil society. Our decisions about what to allow will necessarily be political and negotiated. We will achieve agreement at one time that may give way to revision and a different agreement at other times. That is how democratic processes are supposed to work.

If this were simply a matter of protecting innocent lives—or potential lives, or possible lives—then the decisions might not be so difficult and we could afford to err on the side of protections. But what made George W. Bush so anguished before his decision to allow some federal funding for research on some, though not any new, stem cell lines was the fact that some embryo research offers considerable promise for increasing scientific knowledge in ways that may well lead to significant medical advances. Now we have to weigh the potential advantages for the many against the harm to some embryos that may or may not be considered as full lives. This is why citizens of the United States and many other countries are deeply torn in their views about embryo research. For many, society seems to be poised on the brink of a dangerous precipice of technological power to manipulate human embryos that will distort life itself. They fear that we cannot see what lies beyond, and ought therefore to fear the consequences of wrong decisions and not trust our political and social processes to yield wise moral judgments. Again, this is not a matter of morality versus scientific research, but rather a matter of competing values and contested interpretations of what research can and should do.

Those seeking to protect the presumed interests of embryos must therefore not be allowed automatically and unreflectively to trump those seeking to protect other legitimate interests. Two persuasive girls with juvenile diabetes perhaps put this case best. Eleven-year-old Tessa Wick worries that powerful senators like Sam Brownback are in the position to outlaw the stem cell and therapeutic cloning research that she believes is her best chance at a cure and at her survival. She finds it upsetting that they could choose "a bunch of cells" over her. "It's so scary to me that this guy I don't even know can do that." "It's like he's killing me." Katie Zucker, thirteen years old, agrees. After meeting with an avowedly pro-life senator who from her father's perspective was not interested in any other points of view, Katie said, "As far as I know a skin cell doesn't suffer the way I suffer and a skin cell inserted into an egg cell doesn't have dreams like I have."[3]

Tessa, Katie, and their counterparts have persuaded powerful senators like Utah's Orrin Hatch to modify their heretofore solidly pro-life politics and to support federal funding for stem cell research, for example, because the interests of these real people and their individual lives matter more than what might or might not be considered the lives of embryos. Obviously, different views coexist; they conflict; and they demand decisions, since even no action allows some positions to prevail over others. The question is how to achieve reasonably balanced policies when we do not immediately all agree on the perfect solution.

The history of embryology reveals what is new and what is not new about the decisions that face us today. Perhaps we are on the edge of a political precipice, but only if we do not draw on our considerable past experience and realize that the ground is more level than we may think and that we can learn from past decisions how to make more informed and effective decisions. The cliché is true: only if we fail to learn from the past do we risk making the same mistakes and enduring the same recurring battles. At least since Aristotle's time, in the fourth century B.C., people have argued about when a life begins and how to define a life. Over and over, different groups have adopted a range of interpretations. Some have held that a human life begins from the first moment, whether that was seen as occurring through the human act of intercourse or after the male and female fluids had had time to mingle, or perhaps only at some later point when the soul entered the body or when the vital spirits began to act. For advocates of this view, a life begins sooner rather than later. Typically, a life is seen as fully formed, or at least predetermined or predelineated in some way, at that moment. In contrast, others see life as a continual process of becoming. It takes time and a series of changes to make a life. The process is gradual and emergent, with form and the new life arising progressively through time (epigenetically).

The seventeenth and eighteenth centuries brought a series of well-developed positions on both sides. The most extreme "preformationists" held that there is a moment when a life begins and at that point, the individual is already formed and ready to begin growing. "Epigenesists" held, in contrast, that a life and living processes occur through the course of development, that there is no decisive defining moment but rather an ongoing process, and indeed that is why there is a prolonged course of gestation and development. These positions were well established and provided a set of clear alternatives. Later chapters will lay out in more detail what was at issue and why the positions emerged as they did. The point here is that our current debaters hold the same range of views: a life begins at the beginning, at conception, or alternatively, it emerges only later through a process.

Seeing current debates in light of the past teaches us several things. First, we see that these competing interpretations about when life begins are long-standing and draw on similar arguments. By viewing current claims of moral truth in historical perspective, we can defuse the efficacy of the argument—even if not the passion of the arguer. Second, perhaps we can learn something from past responses to hard questions, though local contextual contingencies will surely shape the particular reactions. Most interestingly, we can understand the way that the past debates have shaped and constrained our current conditions so that they act as what biologists call "developmental constraints." Because of past traditions and the way they have become institutionalized (with extreme preformationist views

WHOSE VIEW OF LIFE?

residing especially in some churches, for example, and the strongest epigenetic views allied with pro-choice advocates for abortion rights), the moral, scientific, and political are inevitably intertwined. Understanding the historical development of the views will help us recognize the reasons for the current range of discordant voices that confronted George W. Bush before his August 9 decision.

This is not, then, a question about when organic matter arises, nor about when something becomes living rather than nonliving. The central question is when a life begins, or when we have something that we want to define as a life on its way to becoming an individual, independent organism; this is what I mean by "a life." When do we have something that has definably begun the processes that we typically assign to life, including growth, differentiation, movement, and a range of independent functions? Clearly deciding which point defines the beginning of an individual life is precisely what is contested today and also what has been contested for centuries. Whose view of life will prevail?

Defining when a life begins draws upon science, as history shows in numerous cases, but it cannot end with science. Biology shows us what happens through all the complex stages of development that make up a life cycle, from formation of the germ cells through fertilization and on to death. Biology shows us at what point an egg cell normally becomes implanted in the mother's uterus, when the heart begins beating, when the embryo is defined as a fetus or that fetus becomes capable of living outside the womb—with varying degrees of support. Biology can give us more and more precise and detailed answers to such questions.

Furthermore, biological research and technological advances can change the answers. Researchers have pushed the time that a fertilized egg can survive in glass dishes to later stages of development in humans. Other researchers have made it possible for younger and younger fetuses to survive with more and more complex life-support technologies. These discoveries change the boundaries of what is and is not possible. They give us a more refined understanding of the organic processes involved in life. Science alone does not, however, define when a life begins. Science does not tell us when along the sequence of embryological stages a developing organism becomes something that we want to call a life.

In our particular form of democracy, such definitions are necessarily social—they are matters of convention. Defining an individual human life must be a negotiation among competing and sometimes hostile claims for truth. The resulting decision should surely be informed by the best available scientific knowledge, not made in ignorance of or in conflict with that knowledge, but having the support of science cannot by itself be enough. As with all social decisions, any definitions on which legal judgments will depend ought to respect the rights and interests of relevant parties; they must address the need to balance duties and obligations toward all, to avoid harms, to protect minorities, and to seek justice consistent with democratic principles as they play out in our complex, pluralistic, and secular society that depends on science for many things.

Yet even if science cannot dictate the answers by itself, any definitions should remain consistent with our best scientific knowledge, so that no political or social decision should rest on false claims about biological "fact." Our conventions should not rest, for example, on erroneous assertions that the fertilized egg in a glass dish is functionally and biologically the same as the implanted embryo, or that an early embryo is biologically functionally equivalent to an eight-month-old fetus, for example. There are threshold stages in development that result in important differences in the capacities of each. This is not, as some bioethicists and many theologians have asserted, "just" a matter of semantics. The differences are real and substantive.

Each stage of a developing individual is alive, but we might well decide that the earliest stages do not yet meaningfully count as a life. Such organic forms do grow and differentiate, but prior to the fetal stages they are not yet formed and do not have their vital organs. They are not miniature persons in any sense, though some bioethicists and some theologians would like to suggest that they are really potential persons and as such have the same moral worth. Any such claim requires a demonstration that potential persons are, in fact, morally equivalent to actual persons, and that we value them equivalently. Any such claim should recognize the vast and significant biological differences between the potential and the actual and then must choose explicitly to discount those differences as not decisive. That is fair. What is not fair is to pretend not to be discounting the real differences, to be relying on biological ignorance or misinformation and suggesting that all stages of human development, from fertilization onward, are effectively equivalent. Doing so requires making auxiliary assumptions that are deeply contested at best.

This book will examine the way these discussions have played out through past centuries and how making such assumptions and thinking about their implications can inform current debates. What were the preformationist and epigenetic positions, and on what reasoning and evidence did each rest? To what extent are these biological or scientific questions, and how far does the science take us in making what are necessarily also philosophical, political, and social decisions? How can we understand how and when to develop the science in a politically contested environment, and how should we go about making wise and informed decisions as we move toward an exciting and sometimes frightening future of ever greater opportunities and challenges?

History can show that we are not on the brink of some new type of danger that we have never encountered before. True, scientists can do many more biological manipulations than they could even a decade ago. But rapid advances in science have occurred many times before, and every time impending doom was predicted, it did not come. Rather, science works by accumulating knowledge, revising interpretations in the light of new evidence, and moving forward through consensus of a scientific community about what is established and what is hypothetical and in need of further testing. Indeed, our ability to accumulate new knowledge and to control life has led to astonishing medical advances. Millions of people benefit every year from the development of surgical procedures, drugs, and new knowledge about how the human body works. Where we see challenges and worries, we should also see a balance of possibilities and opportunities.

As a whole, this book offers an argument for studies of biology and society that bring together the diverse types of expertise and perspectives informed by science, history, philosophy, bioethics, and policy. The challenges are too important to be left to only one perspective or another; all points of view are essential to informing decisions about the biosciences that both help to define and are defined by who we are. We must learn to negotiate wisely among the competing positions, to forge intelligent and effective bioscience policy that moves us beyond intransigent extreme positions. Otherwise we will be doomed to cycles of intuitive and uninformed reactions that cannot be expected to bring about a healthy or wise resolution.

From the Beginning

"In the beginning," as for so much of biology, there was Aristotle. He provides the starting point for the gradualist view of life's beginning. Though he wrote a book entitled *Generation of Animals,* he did not segregate study of the development of a single organism into a separate field of "embryology." Rather, he considered it a natural part of the ongoing, inevitable change in the natural world and therefore part of the wider discussion of all generation and corruption. Generation brings into being; generation of animals and plants involves bringing to life, and for our purposes it is these processes that matter. The particulars that allow reproduction and generation of animals show that life is part of nature, exhibits predictable regularities, and is subject to the guiding processes shared by all sublunary nature (that area of the universe "below" the moon and including earth).

Above all for Aristotle, generation involves change over time. For the generation of individual animals, including humans, the process brings together the "semen," or fluid, from the male and the female and thereby provides the material and motive causes for development of an individual. As with all of Aristotle's physics, four causes (material, formal, efficient, and final) act together to shape the original seminal fluids into a fully formed, adult individual of the right sort. The female contributes the material cause without which there can be no thing. This matter resides in the menstrual blood, Aristotle believed, and after "the discharge is over and most of it has passed off, then what remains begins to take shape as a fetus." This menstrual blood is not pure, however, and cannot do anything by itself. It is "that out of which it generates" and must be acted upon by the male semen, which is "that which generates." The male provides no material cause, but only an active stimulus for the dynamic development that follows.

The formal cause that provides the form and the efficient cause that actually brings the animal into being both act through the coming together of the male and female fluid contributions. Only at that point can the final cause, the telos of the organism, act. These four, then, cause the generation of each individual animal, beginning with the mingling of the male and female "semen" through sexual reproduction. "Thus things are alive in virtue of having in them a share of the male and of the female, and that is why even plants have life." The male and female serve as the "principles of generation."[1] Each individual life begins anew, from Aristotle's view, in a way that seemed to later commentators rather like spontaneous generation.[2]

Yet there is more to understand than that. It is not simply matter, set into motion and guided by form and telos, that will produce an individual life. Nor can it be the case, Aristotle urges, that the form emerges because of some outside cause. No, the causes must be internal. An individual life also requires a "soul" that must be there "from the outset" of life and resides within the material body. This soul guides the gradual unfolding of form (or epigenetic process) by which the living form gradually emerges from unformed material. For Aristotle, the "soul"—consisting of vegetative soul for all living beings, plus locomotory soul for all beings capable of picking themselves up and moving around, plus rational soul for humans alone since only humans have the power to reason—causes the potential of the body to yield an actual life. Or "To answer the question, How exactly is each of the parts formed? We must take first of all as our starting-point this principle. Whatever is formed either by Nature or by human Art, say X, is formed by something which is X *in actuality* out of something which is X potentially." It "is clear both that semen possesses Soul, and that it is Soul, *potentially.*"

In the beginning of any individual animal, then, an external agent sets the process in motion, by bringing together the male and female fluids through sexual reproduction. After that, however, the causes are all internal, including the final cause and attendant soul. These causes lead to formation or differentiation and growth, for "once a thing has been formed, it must of necessity grow."[3] The whole with all its parts does not exist at first, but each part gives rise to other parts, in a process set into motion by the initial act and the continuing working of the causes and guided also by the teleological drive to actualize the relevant potential. A living body differs from mere material, therefore, because of the action of the causes, including the soul.

When referring to what has typically been translated as "soul," Aristotle did not mean anything like a Christian or Jewish soul. With no concept of a Christian deity, he certainly did not envision a spiritual, divine, or somehow supernatural soul. Rather, the soul is that which produces life; it serves as the final cause that actualizes the potential life that is there in the material and in the motion when the male and female contributions come together. Only at a later stage has the growing material acquired enough of the gradually, and sequentially, emerging parts to be considered a miniature being and to have all the causes and soul realized fully.

Aristotle also did not picture an embryo in our sense; he had no reason to believe that any tiny, materially bounded cell exists from the beginning, with conception. This is the main point of his emphasis on the gradual coming into being or the process of generation. He did not know that animals generally begin as eggs, since he could not see them (with the exception of eggs from chickens, frogs, insects, and such). Indeed, most higher organisms' eggs remain so well hidden inside the body that they cannot be seen even by the most aggressive observer without killing the subject. Aristotle wanted to observe generation and life, not death. As a result, in mammals he "recognized" only the unformed fluids apparent during and after sexual intercourse. What Aristotle saw seemed manifestly unformed and hence obviously form must emerge gradually from unformed material. Yet, with time and additional observations and philosophical reflection, other interpretations emerged to challenge Aristotle's.

Religious Interpretations

In the absence of more direct empirical evidence about the beginning of a life, the matter remained open to interpretation by philosophers and theologians. Where clear statements are not available, we can often gain a good sense of prevailing opinion by looking at legal rulings and interpretations. There is a rich literature on religious interpretations of embryos, so all we need here is a brief introduction to the most relevant considerations.

Early Jewish law, as Laurie Zoloth summarizes nicely with direct reference to current discussions of cloning and stem cell research, is consistent and clear. For forty days, the fluid of intercourse that will become a fetus and an adult lacks "humanity," is "like water," and is not yet a "person."[4] The Torah makes an important distinction that gives us clues about the early views of embryos and their implications. In the first forty days, someone who causes a woman to miscarry is judged to owe monetary damages only; after that time the crime is ruled to be murder punishable by death. As Zoloth points out, this is not a straightforward matter of clear moral distinction, however, since Jewish law is complex, with different categories of prohibitions and moral value. An action may be prohibited on secular and pragmatic grounds, even though it is acceptable on religious grounds. Nonetheless, at forty days ensoulment apparently occurs and the Jewish moral and legal status of the embryo changes.

Early Catholic interpreters followed primarily Aristotelian accounts. While Saint Augustine is widely cited for having condemned abortion as breaking the moral connection between sex and procreation, he also wrote that "who is not rather disposed to think that unformed fetuses perish like seeds which have not fructified?"[5] Becoming a human life occurs through a process that results in a human only later, after the fetus has begun to grow. Though abortion is a sin, it is not homicide since hominization has not yet occurred. Various individual legal rulings upheld the same interpretation throughout the Middle Ages.

Saint Thomas Aquinas agreed. He opposed abortion as against marriage and the moral duty to procreate. Yet he did not see abortion as murder before the embryo was ensouled and had become human. For Aquinas, the fetus first acquires a vegetative soul and begins to grow. Then comes an animal soul and finally, only when the body is developed with all its parts present, a rational soul. This interpretation of "delayed hominization" prevailed for most of early Catholic history.

Only in 1588 did Pope Sixtus V issue the first papal declaration on this subject. Entitled *Effraenatam,* or *Without Restraint,* his statement was intended to confront the rising prostitution in Rome. Pope Sixtus V decreed—though he did not declare infallible—that contraception and abortion at any stage were considered homicide. Both were subject to the highest spiritual penalty, excommunication. Only three years later, Pope Sixtus died and his successor, Pope Gregory XIV, rejected the interpretation and penalty as excessive. In particular, Gregory regarded the ruling as inconsistent with the theological understanding of ensoulment. He may also have found the sheer volume of cases overwhelming.[6]

Islamic interpretations have been more open; most credit a human life as beginning with ensoulment at forty days. Muslim law accorded with that interpretation. Because Muslim practice is more like the early Catholic practice, in that multiple coexisting and competing interpretations are allowed, there were and continue to be differences of opinion. Some Muslim scholars regard ensoulment and hence human life as beginning around 120 days, while a few argue for earlier stages. Again, we see this played out in legal rulings: causing a fetus to die before forty days has not quite been considered homicide, as it is after that time.

In the absence of evidence from biological investigation, such differences of opinion continued, but the dominant view was that a human life is ensouled and therefore begins with "quickening," or the "coming-into-life" that was thought to occur around forty days, which is roughly when mothers begin to be able to detect changes. The rise of empirical study began to change that assumption.

Seeing Inside

The sixteenth and seventeenth centuries brought renewed enthusiasm for empirical study of nature, stimulated by what is known as the Scientific Revolution. In addition to studying the effects of matter in motion and examining the heavens with new instruments, researchers became intrigued by the possibility of "seeing" inside the living organism. Leonardo da Vinci had led the way, but secretly and therefore without immediate followers, in his unpublished observations of fetuses and female reproductive parts. He made the first-ever, astonishing sketch of a fetus curled up in the womb as if comfortably sleeping and not quite ready to be awakened to life outside its cozy environment. In 1521, Leonardo devoted an entire notebook to such sketches.[7] (See Figure 1.) Since his marvelous studies and interpretations remained locked away, however, they did not contribute directly to the generation of new knowledge, nor to new approaches to embryology. Only much later could we admire the beauty and be astonished by Leonardo's embryological representations.

More generally, medical anatomy did contribute. As Italian anatomist Andreas Vesalius and early-sixteenth-century artists stripped away the outer layers of the living body to expose the underlying structures and to suggest the functional connections within, they inspired other anatomists and artists to look inside. But how could they look inside a mother to see the very beginnings of new life? Not only were the structures tiny, but specimens were relatively rare. Every body has a heart, for example, but only those few females who are pregnant have embryos or

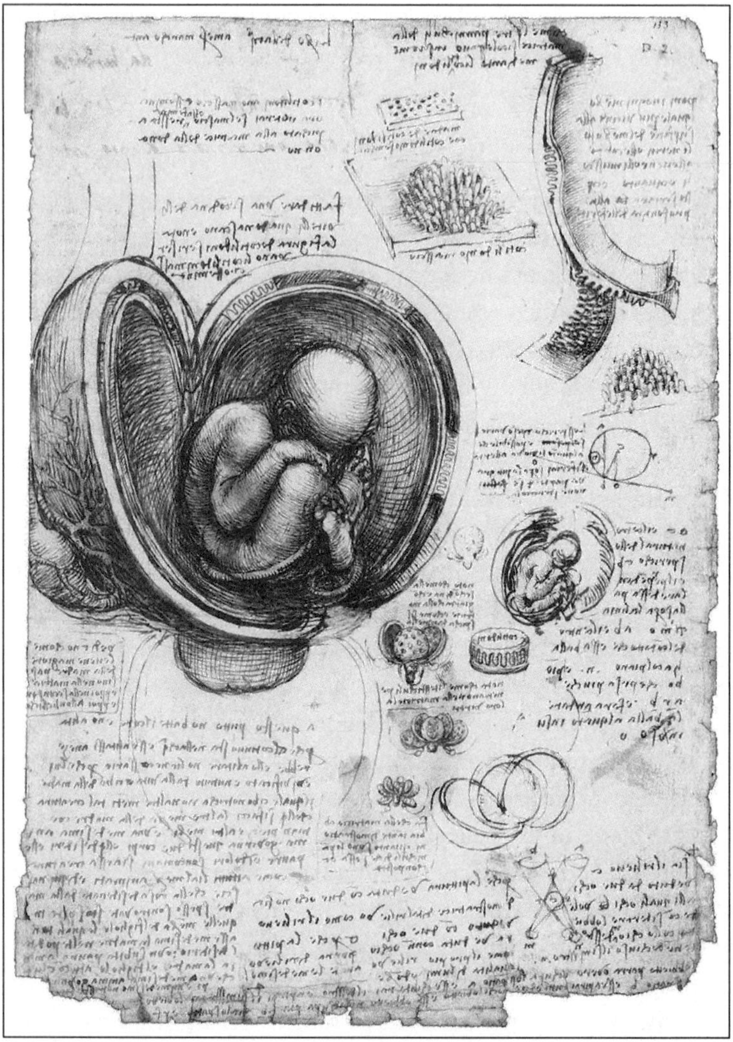

Figure 1. Leonardo da Vinci's notebook sketches of human fetuses. Reproduced by gracious permission of Her Majesty Queen Elizabeth II, copyright reserved. The Royal Collection © 2003, Her Majesty Queen Elizabeth II.

fetuses. Microscopes were available to help see inside, but only in cases where there was something to look at.

Since living organisms are obviously not transparent, literally looking inside a body unfortunately demands that the life under investigation be killed. As a result, studying the internal processes of living organisms requires epistemological assumptions about what will count as knowledge of life from dead specimens and methodological assumptions about how to get that knowledge. Looking inside mothers to see the developing offspring—especially for humans—might provide information but was only rarely possible. This left theory as an option for imagining how the commingling and coagulating of male and female fluids might work, while others strained hard to interpret what they could actually see. Still others sought new ways to "see."

One such approach is to look at other animals and make the assumption that if they look like humans, then during early periods of development they should presumably follow the same patterns that occur in humans. Since killing and dissecting other animals seemed easier and more acceptable than working with human subjects, comparative anatomical study was taken as providing something of a window into the deep recesses inside a human mother. Researchers began to gain a sense of patterns and processes of development.

Mapping out developmental sequences in other animals at least provided a starting point. "Monsters" (or teratological specimens) and dead fetuses also yielded clues, though again only with the additional assumption that the dead material retains the form, even if obviously not the function, of the living organism. If so, then a dead fetus expelled from a mother or lying inside a dead mother offers evidence about what normal human fetuses look like. Museums, medical collections, and private "cabinets" accumulated whatever aborted fetuses they could find, including those collected from mothers who had died. These specimens preserved in jars still fascinate modern viewers, but the effect must have been considerably more affecting in the past.

The impression of a dead human fetus, preserved in its interrupted developmental stage, must have been startling and a bit horrifying in a time before cameras, not to mention sonograms, ultrasounds, and other modern viewing tools. Today, expectant mothers with good medical insurance can have copies of pictures and even movies of their fetuses at various early developmental stages. For centuries, however, they had little sense of the process going on inside until quickening—or morning sickness and other physical changes such as suspension of the menstrual cycle—provided the clues. The collections of often distorted aborted fetuses floating in glass jars still seem somewhat macabre and removed from real life, even while they hold special poignancy because of their fragility and failure, representing both beginning and end, life and death.

Artists may draw what they see or perhaps elaborate to interpret beyond the visible so far as their imaginations take them. Representations of female reproductive anatomy often revealed more imagination than accurate reflection of the visible, especially during the centuries when dissection was prohibited, restricted, or just not standard practice and therefore not available to inform the artist or anatomist. By the eighteenth century, sculpture and wax models joined the stock of two-dimensional images of the human body. What took Vesalius a number of separate paintings to show, as layers of the body were peeled away, anatomical models could illustrate in one three-dimensional specimen. For example, the wax models at La Specola in Florence include a number of fetal models and one marvelous woman. With realistic hair and a strand of pearls, the woman looks so real lying there naked. As observers begin to remove the wax layers of outer skin, then layers of organs, eventually reaching the tiny fetus hidden there, they can peer inside in a way that anatomical study of fragile tissue does not allow.[8]

No matter how beautiful the models and notwithstanding the attention to detail, artists and anatomists could still depict only what they could see or imagine. The collection of fetuses and specimens of deceased pregnant women available for dissection fortunately remained limited and therefore offered too little information for artists to document development over time. Even large collections could present only a compilation of widely separate "snapshots," nothing even close to a virtual "motion picture" of the entire developmental process. No collections documented the very earliest periods of development, before something recognizable as a human form became visible. How, after all, could people collect and present something they did not even know was there? Despite the best intentions and the most passionate interest, researchers simply did not know how to look inside the pregnant female at those earliest stages when it was not yet even clear that she was pregnant, nor to follow the processes of development over time.

Without knowledge of the earliest parts of the developmental process, researchers had little reason to move beyond Aristotle's interpretation of life's beginning, that two fluids come together and mix. Gradually, however, the situation began to change. William Harvey, best known for his studies of the heart and blood circulation, also studied the generation of animals. For his *Exercitationes de generatione animalium* in 1651, he dissected female deer shortly after copulation and also examined incubated chick eggs.[9] The does revealed no eggs, no female semen, and no evidence of a mixing of male and female seminal fluids. Instead, Harvey was convinced that "Omne vivum ex ovo" (each organism arises from an egg) and that furthermore the eggs arise as a product of conception. His study of chick eggs further persuaded him that the chick arises gradually, epigenetically, with each part arising in order and not all at once. David Bainbridge reflects on Harvey's contributions to understanding human development in his *Making Babies,* which also demonstrates the gradualness and continuing interactiveness of development.[10] Harvey's epigenetic view of generation, like Aristotle's, fit what he saw.

We must not overinterpret the work of early thinkers, however, nor impose our own images on their interpretations. By "egg," Harvey did not mean what we mean. Yes, he could see the chick egg perfectly well. But, as historian Richard Westfall pointed out, what Harvey meant by "egg" more generally was highly ambiguous. Apparently, what he saw as an egg in the deer was actually the amniotic sac, and in insects it was the butterfly's cocoon. What Harvey meant was that there is some defined, localized material primordium for each organism that arises from conception and provides the starting point for the gradual emergence of that individual animal. He gave a common interpretation, which brings all animals and humans together in the same generalization about development: "An egg is the common origin of all animals."[11] These eggs are unformed, and for all practical purposes homogeneous. A "formative virtue" guides development and produces form from the unformed. Harvey's epigenesis therefore accorded nicely with his vitalistic view of life, according to which life consists of matter in motion and a vital something that makes it alive.

The first *Encyclopaedia Britannica* (1771) reflected epigenetic thinking tracing back to Aristotle. This volume defined an *embrio* as "the first rudiments of an animal in the womb, before the several members are distinctly formed; after which period it is denominated a foetus."[12] The word is said to have come from a combination of *em* and *bryein,* for "swelling inside." As historian Shirley Roe has detailed, the epigenetic interpretation was countered by an alternative, the preformationist view, that attracted equally ardent supporters during the eighteenth century. They offered competing understandings of what that internal "swelling" really involves.[13]

The epigenesists, like Aristotle, all emphasized the gradual coming-into-being of form. Indeed, many pointed out, just by looking at early developmental stages, empirical naturalists could see that all the parts were not yet fully there. Form must, therefore, emerge gradually through a process of becoming. A sort of "seeing is believing" argument prevailed—or more accurately "not seeing is believing not." But this formation process, and the fact that it acts in a consistent and predictable way, presumes some guide to the emerging form. Most popularly, that guide was taken to be a vital force or property of some sort, perhaps helped along by or perhaps independent of Christian "ensoulment."

Being an epigenesist did not require also being a vitalist (or invoking special vital causes to explain life) in the seventeenth century: René Descartes and Pierre Gassendi provide examples of materialistic epigenetic views. Descartes suggested that the male and female semen come together and "ferment," while Gassendi offered a more physically concrete suggestion that each individual begins when particles come together from both parents to make a seed, out of which form gradually arises through a mechanical series of unfoldings.[14] Yet as Roe demonstrates, aside from these two, during this time epigenesis was usually associated with some form of vitalism. That philosophical assumption became an increasing problem during the eighteenth century, as vitalism itself came under scrutiny and its proponents confronted demands for further explanation.

The most important alternative to vitalism came from materialistic preformationists. It was not so much that they started with preformationism, a conviction that all body parts exist from the beginning of the organism, preformed and ready to grow. Rather, materialists or mechanists began by seeking to banish vital forces or entities from science and to account for life, as for all of nature, in terms of matter in motion. Since explaining the generation of form from unformed, effectively homogeneous matter seemed to require some vitalistic and often teleological cause that was hence not strictly material, and since this was *a priori* unacceptable to materialists, they arrived at the conviction that form must be there from the beginning. Building the form in from the beginning had the considerable advantage, therefore, of providing an explanation without invoking nonmaterialistic causes. The preformationists were so guided by their grounding in materialistic assumptions that they accepted the necessity of the form's existence even when they could not see it. Not seeing should not necessarily lead to not believing, in other words; for preformationists empiricism cannot provide reliable knowledge.

Preformationists also rejected what was *de facto* a sort of spontaneous generation in Aristotelian epigenesis, where each generation brings new life and form into being out of unformed nonlife. There was just too much change without apparent material cause to provide what they could count as a proper explanation. In contrast, preformationists did not allow each individual living organism to come into existence *de novo* but rather only in continuity with previous form. This might happen as one generation gave rise to the next one, but the extreme logical form of preformationism sees every future individual as "encapsulated" in the very first generation of that type of organism. Nicolas de Malebranche's *emboitement* ("encasement") invoked images of tiny little beings packed away inside the first humans —all the way back to the beginning.

Among preformationists of this sort, or those who advocated preexistence, spermists held that form resides in and is passed on through the sperm; ovists maintained that the female contribution carried the form, presumably through the egg.[15] With a materialist frame of mind that called for accounts in terms of physical matter in motion, Nicolaas Hartsoeker in the late seventeenth century performed studies with optical instruments that led to one of the most extreme versions of this interpretation. Hartsoeker provided a striking image of a homunculus (see Figure 2), a tiny human folded up inside the spermatozoa—the male contribution that Anton van Leeuwenhoek discovered with his single-lens microscope. Reproduction somehow activates the homunculus that has been passed along from the father. In its more extreme and clearest version, each generation begins when the process of reproduction awakens the tiny form and begins it on its developmental course to birth. Each form comes from the previous generation, back to the first human. For extreme ovists, for example, all future humans started out in Eve's eggs. No spontaneous generation here. No mysterious life arising from nonlife. Each individual life can only arise from a preexisting life form, though a process must occur in each individual to bring about its full development and growth.

Others offered more moderate views, grounded much more solidly in microscopic observations. Historians and contemporaries have widely regarded Marcello Malpighi as the best embryologist of the seventeenth century. With careful technique, he took freshly laid chick eggs, opened them, removed the develop-

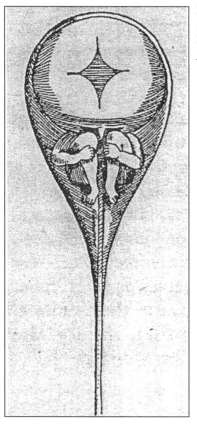

Figure 2. Nicolaas Hartsoeker's preformationist homunculus. From Hartsoeker, *Essai de Dioptrique* (Paris: Jean Anisson, 1694).

ing chick from the surrounding nutritive material and laid it out on glass. This allowed him to see the tiny parts of the emerging chick, and he concluded that the embryo is formed at the time of fertilization. After that, development involves that form's becoming visible. The way it looks to us changes because the form becomes more visible, not because the form changes from nothing into a formed something. As Malpighi put it, "while . . . we are studying attentively the genesis of animals from the egg, lo!

WHOSE VIEW OF LIFE?

in the egg itself we behold the animal already almost formed, and our labor is thus rendered fruitless. For, being unable to detect the first origins, we are forced to await the manifestation of the parts as they successively come to view."[16]

Among contributors to the discussions, the contrast between the German embryologist Caspar Friedrich Wolff (in 1759) and Swiss naturalist Charles Bonnet (in 1769) brings the competing eighteenth-century views into sharp contrast. Both studied chick development. Each saw, for example, the twenty-eight-hour stage, shortly before the heart becomes clearly visible and is clearly beating. Given that they had no way to watch the process every instant, they had to take what in effect were snapshots and extrapolate to assumptions about what happens in between. Wolff, the epigenesist, looked and did not see a normal heart beating at twenty-eight hours. He concluded that it had not yet fully formed. If he could not see it, then there was no legitimate reason to assume it was there. Rather, he concluded that it was not there and only came into being over time, through the causal agency of the *vis essentialis* ("vital force").

Bonnet did not see the formed heart either, nor any beating. He did not claim that he did. Rather, he concluded that it must be there. It had to be for a rational, materialistic explanation of the later form to make sense. It must be limitations in our ability to see that prohibit us from seeing what must be there. Advances in microscopy had already revealed so much more than we could see before, that it was quite reasonable to assume we would be able to see even more with better instruments. Stars, cells, microscopic mites, internal structural details: just imagine how much more would be visible in the future. Not seeing should not lead to not believing.

Others agreed with Bonnet that all parts were there all along, but simply transparent and too small for us to see. As Albrecht von Haller put it, "The manner by which these same parts become visible from being invisible is of the most grand simplicity. It is the effect of growth, but even more the simple effect of opacity."[17] For von Haller and other preformationists, accepting the existence of invisible transparent form made more sense than appealing to the postulated existence of invisible and by its nature inevitably unobservable vital forces or entities. At least from their view, better equipment could, with time and technical progress, make the form visible. These debates, sometimes passionate and hostile, persisted even in the absence of decisive data or compelling independent preferences for one or the other point of view.[18] They set the stage for similar competing positions and debates today.

At its heart, the debate was about whose view of life would prevail concerning when an individual life really can be defined as beginning. For epigenesists, the form and by implication also the life of the embryo and then the organism emerges only gradually. There is no chicken there in the egg, but only the background conditions that will make it possible for a chicken to come into being if a number of other conditions are also satisfied. Since a life comes into being by stages, it necessarily remains a matter of interpretation at which stage we can say, "Aha! A new life has begun." Yes, this occurs at some point rather than another. But deciding which point is a matter of definition and convention—it is not a matter to be dictated by some deep truth lying within. The definitions should be informed by and consistent with the best available scientific knowledge, but they require making choices and judgments beyond the science itself. When critics today complain about attempts to distinguish "preembryos," "preimplantation embryos," and embryos, for example, they insist that the effort is mere semantics and misses the real point. What we see from this brief history of past debates is that it is indeed a matter of words and definitions, and that distinguishing differences by the careful use of terminology is precisely what we need most to make wise decisions and informed judgments.

In contrast, preformationists or predeterminists who hold that the life and its form are laid down in the beginning or with fertilization have problems with the details of embryological studies. For them, embryos are literally preformed, tiny material beings that swell and grow larger in the course of individual development. They are defined from the beginning. It is easiest to assert that each is also fully alive and already an individual life. This interpretation points to the coming together of two cells and their genetic material as the beginning of life. The preformist interpretation, favored by many conservative religious groups today, lends itself most readily to a strictly genetic determinist view. If the coming together of material and inherited cells and nuclei is decisive, then apparently genes make the man. And if it is something else that is decisive, such as a soul or a vital something-or-other, then we are back to conventional views about when life begins. For there is no logical or material reason that this soul or other factor need start the life at fertilization.

This brings us to religious interpretations. For humans, which is what people naturally enough care about most, the standard view remained that of the traditional Jewish, Muslim, and even Catholic theologians: an individual human life begins with "ensoulment." At this point, thought to be forty days after conception, some mothers can first begin to detect their pregnancy. There is, indeed, an embryo with a basic human body plan in place. The process of generating a new life has begun. It becomes a life because of the process of generation and the action of causes (for Aristotle), ensoulment (for traditional religious interpretations), action of the vital forces or entities (for vitalists), or attainment of a certain stage of growth (for materialists). These basic perspectives remained, even as observers added new descriptions and researchers developed more detailed interpretations.

The Mammalian Egg

The picture changed, however, with Karl Ernst von Baer's discovery in 1827 that even mammals start as eggs, just as do chicks, frogs, and other species whose eggs we can see so easily. This is not Harvey's hypothetical "egg" that he assumed must be the material beginning of any new life. This was a tangible, visible, material beginning. Primed to look for it, von Baer first observed this egg in a dog belonging to his physiology professor, Karl Friedrich Burdach, in Dorpat, Russia. Burdach's willingness to sacrifice his own dog surely earns him some recognition for this great discovery as well. Not only did von Baer see the egg, but he also determined that it seemed likely to have been passed on by the mother rather than having arisen by spontaneous generation or *de novo* because of fertilization. As he wrote at the end of his study of mammalian eggs, "Every animal that originates from the copulation of a male and female develops from an egg, and none from the pure 'formative' liquid."[19] By this point, the competing epigenetic and preformationist positions had settled into entrenched differences; an empirical discovery alone could not dislodge adherents from their respective certainties. As in today's discussions, the debate was less about science and what could be observed than about other, nonscientific assumptions. Yet this discovery of the mammalian egg provoked new questions and approaches, and in the long run it did make a significant difference.

If, indeed, each individual life begins as an egg, or at least if the egg exists at an early stage, as von Baer showed it did for mammals, then there is material continuity across generations. At least embryologists must ask about the nature of the egg and to what extent it carries something with it from parent to offspring. New life might arise through an epigenetic process, as Aristotle had been so sure it does. It might even come into life from nonlife at some point, and certainly there comes a point when it is independent and alive. That moment of generation and the processes that trigger it must be understood in terms of the material egg, with its continuity to the mother.

We often focus on the great leaps of theory and interpretation that carry us forward to new understanding. We typically do not see as easily the painstaking accumulation of data or consider advancement of techniques and equipment as important. Perhaps we are trained by thinking about "scientific revolutions" to award more credit for creative insight and hence more value to theory and interpretation. The earth goes around the sun, which is not the center of the universe: in retrospect this heliocentric world view seemed so radically different from the previous interpretation that our attention remains focused on the theoretical achievements of Copernicus, Galileo, and others. We know, of course, that Galileo turned his telescope to the heavens and made important discoveries, but the general public is much less likely to have a sense of Galileo's contributions to our view of Jupiter's moons or lunar craters than to know what is meant by "the Copernican revolution." Similarly, we know that Darwin gave us a theory of evolution by natural selection, yet the firmly grounded results of his studies of fossils or pigeon breeding are much less well known. Nonetheless, even these examples of the practice of science are more familiar to general readers than are the details of embryology.

For studying life, advances in microscopy were absolutely necessary to move beyond an Aristotelian picture of fluids vaguely mixing together. The single-lens microscope made it possible for Leeuwenhoek to see spermatozoa, for example (though he concluded that the spermatozoa were parasites). Leeuwenhoek believed that he could see tiny animals, or "animalcules," that mated and gave rise to animals, an interpretation that soon lost favor in the face of accumulating evidence from stronger microscopes and advanced techniques. He asserted that spermatozoa enter into and join the egg, but he did not observe the union clearly and the claim was hotly contested. Whereas Leeuwenhoek relied on the single lens of a simple microscope, others used the dual lenses of the complex microscope. At first, the advantages remained limited; some observers could even see better with the single lens because of distortions in a compound apparatus. By the 1820s and 1830s, however, improvements in the lenses had reduced aberrations and expanded the effective range of magnification. With improved lenses also came improved methods of preparing and preserving materials for study.[20]

Von Baer and others used techniques that revealed more clearly details of the newly emerging life. Von Baer watched a chick egg develop, as Aristotle and many others had, and he saw so much more than his predecessors because he used better microscopes and advanced techniques for hardening and "fixing" the egg. A normal egg, as anybody who has ever cooked a traditional breakfast knows, is highly fluid and malleable. Mauro Rusconi made the valuable discovery that boiling eggs at various stages of development preserves the details of embryonic development. This fixing process allowed observers, in effect, to halt the process of development at a particular moment and to examine slices of different developmental stages. They assumed that what they saw would yield valid knowledge about the normal life processes. Von Baer developed this approach further and used the capabilities of improved microscopes to significant advantage. Working his way through the hardened egg, he looked first at the outer surface and then peeled away the layers to look inside. He wrote of earlier researchers that they could not see the details "doubtless because they were not aware of any method for removing the white of the egg in order to harden the yolk and thus analyse it."[21]

Chick eggs were especially useful to early embryologists since they are plentiful, fresh year-round, relatively inexpensive, and reasonably forgiving of being poked, prodded, and even cooked. Frog eggs are also large and forgiving, and they develop entirely outside the body with no opaque shell. Different species of frogs have different colors and sizes of eggs, as well, allowing useful comparison. In 1834, von Baer described the earliest stages in frog development. The egg develops a furrow, at first shallow and then deeper, eventually dividing the egg into two parts, then four, and so on. He followed the process to the tenth division in brown frogs (see Figure 3). Von Baer beautifully depicted these divisions and described the regularities in the process, which he saw as disproving preformationism.[22] How, after all, could a tiny, preformed being exist in the unfertilized egg, be divided up by such furrowing, and retain its form? This made no sense, von Baer insisted. Yet he was not sure just what structural and functional significance the furrowing and segmentation of parts had. Was the process perhaps designed to allow each separate part to have access to the sperm and to advance the causes of epigenetic development? Why this segmentation?

The newly introduced cell theory provided one framework for interpretation. Yet it took until the late nineteenth century to untangle the embryo's developmental complexities and understand this development in terms of cells. It took even longer to understand why a cellular interpretation of life helped address many

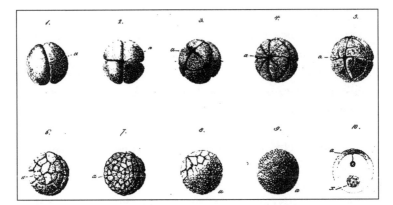

Figure 3. Karl Ernst von Baer's representation of the segmenting frog egg. From von Baer, "Die Metamorphose des Eies der Batrachier vor der Erscheinung des Embryo," *Müller's Archiv für Anatomie, Physiologie und Wissenschaftliche Medizin* 1834: 481–509, plate 1.

key biological problems. It is worth following this trail of discovery and interpretation, as the path tells us much about how researchers made sense of their findings about the complexities of development and the choices they made along the way.

Cells

Soon after von Baer's studies, Martin Barry focused attention on the nucleus and on cells as key to understanding embryogenesis.[23] Barry had studied with Theodor Schwann in Germany. One of the proponents of cell theory, Schwann held that an understanding of cells was a generalizing framework for the study of life, starting with the interpretation that cells are the fundamental units of living organisms. When Schwann and Matthias Schleiden brought together their respective studies of animal and plant cells and placed their observations in the context of cell de-

WHOSE VIEW OF LIFE?

velopment, they provided a framework in which to think about developing eggs.[24] Yet they did not make precisely that step themselves. Instead, they offered a more complicated picture.

Trying to draw generalized conclusions from what was known about a wide variety of cells in the 1830s was already a challenging goal. Indeed, believing that any generalizations about life processes and structures would hold across plant and animal species involved a major set of assumptions. Schleiden and Schwann went further. Seeing similarities in form and sharing assumptions about the unity of nature, they were convinced that not only do plants and animals have the same basic cellular units, but that those cells behave in fundamentally the same ways. They believed cells have the capacity to coagulate or coalesce out of noncellular material.

Schleiden and Schwann believed that they saw coalescence occurring, and indeed it is very easy to "see" if the observer starts with the assumption that such an event is possible and knows what to look for. But that assumption is not well founded. Microscopic observations require considerable skill and interpretation; one needs to know what to ignore as well as what to look for. Someone who has never looked through a microscope before will find it difficult to identify cells and decide where one cell ends and another begins. Yes, there are some shapes that look like cell membranes surrounding liquid internal contents. If we know that is what we are looking for, we can see it. But cells are three-dimensional structures, and the traditional binocular microscope can focus only at one distance at a time, presenting more or less a two-dimensional picture. In addition, the light is probably uneven, resulting in imperfect reflections. Untangling the various layers takes work, and without additional training or evidence of what is going on, the novice will find it easy to misinterpret. Almost surely this explains the occasional parts of Schleiden's and Schwann's reports that seem so odd today.

To Schleiden and Schwann, cells sometimes seemed to divide and to proliferate through this division into a larger number of cells. Sometimes it appeared that they were crystallizing and spontaneously generating out of the surrounding material. As Schwann put it, following Schleiden's initial suggestion, there is "in the first instance, a *structureless* substance [cytoblastema] present, which is sometimes quite fluid, at others more or less gelatinous. This substance possesses within itself, in a greater or lesser measure according to its chemical qualities and the degree of its vitality, a capacity to occasion the production of cells."[25] Schleiden pointed to cell nuclei, which he called "cytoblasts" or the "stems" from which cells arise. Schwann agreed with respect to animals. He painted a picture of a dark granule that he called a nucleolus, which gave rise to the "nucleus" and around which layers of material accumulated to make up a whole cell. The nucleus thus served as starting point and center for both. Yet for Schwann, the nucleus was at times a hypothetical structure that he believed must be there, rather than a clearly defined body visible inside every cell.

In other words, the founders of the cell theory did not really understand or present us with what we think of today as the cell theory. For Schleiden and Schwann, cells were indeed the fundamental units of life. But they could arise out of structureless fluid, as pearls accumulating around a sort of "nucleus." This was a perfectly reasonable interpretation at the time. What they provided was an entirely material and physical interpretation of life, without invoking vitalistic or teleological causes to stimulate the life in each cell. Life could arise from nonlife in each cell, but the process of adding all the cells builds up a life in a purely physical way. This was a reasonable foundation for a blend of materialism and gradualism, or epigenesis. Yet while the cell-as-fundamental-unit-of-life part of the theory remained intact and held up under additional investigation, the interpretation of how cells arise did not.

By the 1850s, culminating in the work of Robert Remak and Rudolf Virchow, the prevailing view was that cells come only from other cells, or "omnis cellula a cellula" as both these men famously put it.[26] Remak first, then Virchow, stated repeatedly and persuasively that cells divide by splitting. Those furrows that von Baer had pictured so clearly are actually dividing the cell into two cells, then four, and so on. Furthermore, it is not just that cells can divide, which Schwann and Schleiden had certainly acknowledged as well. Rather, Remak was convinced that cell division is the only way by which new cells can arise. Remak explained why others might think they were seeing otherwise but argued that each new cell could be explained in terms of cell division.

Virchow, the much more influential cytologist and pathologist, articulated the same view: that all cells can and do come only from other cells. After 1855, the earlier idea of spontaneous generation had given way to seeing life and development in terms of continuity of cells from other cells—through cell division.

This brings us back to Martin Barry and others looking at the nucleus of the cell. It was not clear what role the nucleus plays, but if the furrows were cleavages running the depth of the cell and if every cell has a nucleus, then the cleavage must also cut the nucleus in half and must therefore be a specialized case of cell division. It therefore cannot itself be the direct cause of the cell division, as some had suggested. Barry was not sure how the divisions occurred nor how nuclear division gave rise to new life in the form of the developing embryo, but he noted that "The origin of the embryo from the nucleus of a cell may assist to solve a question on which, I believe, physiologists are not agreed."[27] Remak was certain that the cell nucleus divides when the cell does.

Others were less certain, and there was little understanding of what nuclear division might mean. The focus remained on cells. For purposes of individual development, it was accepted that the egg is a cell. They held that the egg cell undergoes cleavage into more and more cells. Somehow, along the way, those cells become differentiated and bring about epigenetic development of form out of an initially unformed egg cell. Yet this was not the old-style epigenesis, with form arising from the unformed and life arising from nonlife. Instead, the egg cell carried the life along with it from "the beginning," even as the form emerged gradually along with the sequence of cell divisions.

Life came from life, in this view, and never from nonlife. There was no spontaneous generation of life from nonlife, no need for vitalistic introduction of new life, since the life came from the previous generation of cells. Cell division provided the continuity. In turn, the concept of continuity provided the foundation for major advances in understanding the development of individual living organisms. The egg is a cell that undergoes cleavage into more and more cells during development. Thus, seeing the egg as a cell helped make sense of the egg and also reinforced the emphasis on cells as the fundamental units of life. This combination provided a powerful foundation for further scientific study of what constitutes life. There were rich possibilities for other researchers to join in and begin to record and analyze in detail every step, from the egg's beginning as a single cell, through its divisions into more cells, to the cellular differentiation that makes possible the complexity of living organisms.

This cell theory also made possible the identification and study of embryos, through embryology. An egg is a cell. That cell divides and differentiates. It becomes—at some point not definitively specified—an embryo, or a prospective but not yet actually living or actually human being. What, then, is the nature of the cell and of cell division? What makes cells differentiate, and how do all the separate cells in an embryo manage to divide and grow in just the right way so as to bring about the right sort of form, in the right sequence, at the right time? This cytological work provided a new and lively set of research possibilities that did not depend on assumptions about whether we should believe what we see or not. Rather, the cell theory provided a solid material grounding for study of embryos—and of the emergence of an individual life from an unformed, though living, egg cell.

More empirical evidence about eggs quickly showed that the egg provides material continuity between the mother and offspring. But what does the male contribute? Aristotle had realized that both parents must participate in the making of offspring. Yet the egg comes only from the mother. What is the role of those spermatozoa that Leeuwenhoek had first seen? Fortunately, improved microscopic techniques and equipment helped once again. By the 1840s, researchers had observed the presence of sperm within the egg after fertilization. Pursuing the hypothesis that it is sperm that fertilize eggs and actually initiate individual development, several researchers had already discovered through experimentation that eggs with no exposure to sperm do not develop and differentiate. Yet, even showing that sperm must be present or seeing that they occasionally are there does not reveal much about what they actually do. Generally, researchers interpreted the sperm as providing a stimulus that, in effect, excited the egg to begin developing. The sperm must act as a physical, or perhaps a chemical, prod to action. George

Newport's observations proved particularly compelling, as he described the spermatozoa's passing through the gelatinous coating and the vitelline membrane that surrounds the frog's eggs and into the yolk.[28]

Evidently, Pope Pius IX did not need to know more about reproduction and development to proclaim the Catholic Church's official position on the beginning of life. In 1869, the Pope determined, implicitly, that life begins at conception. By this he evidently meant at fertilization, or the point when the male and female material contributions come together, even though precise understanding of what happens became clear only in the next few decades. What Pope Pius IX actually did was to reiterate the understanding that abortion is a sin punishable by excommunication, but to remove the possibility that the early embryo is not a life. This view holds, he wrote in his *Apostolicae Sedis,* that all abortion is homicide. Implicitly, he endorsed the interpretation that life begins at conception, or that hominization is immediate.

The 1870s brought new observations and new interpretations. Indeed, further improvements in microscopic lenses made it still easier to avoid aberrations and to get clearer vision of what was happening inside the cells.[29] The rush of new ideas and new questions also encouraged researchers to make major innovations in ways to see inside cells during the 1870s and 1880s. Within a few years, they learned to stop the cells in their tracks, or to "fix" them in the midst of a wide range of active processes, so that they stopped changing and were effectively frozen in time. Once they could preserve cells, they could slice them into thin sections for easier viewing. Microtomes, sharp knives mounted in a stable base, made it possible to slice through an entire three-dimensional egg or embryo in neat and thin serial sections, much as a deli slicer slices through meat.

Imagine your morning eggs just out of the chicken, or an un-

WHOSE VIEW OF LIFE?

baked pumpkin pie if you like that better. Both are soft, runny, and hard to cut neatly. If you want to cut either, you need to figure out ways to preserve and solidify the samples without destroying or distorting them. Cooking works but may change the structures you are trying to study, so it is important to try other fixatives as well. Then you will want the sharpest knife available to make thin slices and reveal the most detail. And you want the sections as uniform in thickness as possible. In addition, you want the end result to complete an array of pieces in as nearly normal a condition as possible. Cytologists, who wanted to study all sorts of cells in as much detail as possible, worked to develop these techniques and the equipment necessary to carry out the best slicing and dicing and the most effective fixing and preserving. They also experimented with staining, to highlight those parts of the cells that took a stain and thereby to differentiate them from those that did not. The flood of new techniques came hand-in-hand with a flood of discoveries, though it was only by the end of the nineteenth century that researchers developed handbooks and sought to regularize techniques and equipment.[30]

Among those looking at cells were some who focused on the cell nuclei in development. Oscar Hertwig, Hermann Fol, and other expert microscopists found many cases of just-fertilized egg cells with two nuclei in them. Furthermore, they showed that this occurs only after, and apparently because of, the coming together of the male and female contribution through fertilization. Fertilization now meant more than just the commingling of unspecified material fluid or some vague "influence" or stimulus from each parent. Instead, these microscopists showed that each parent contributes a nucleus to the fertilized egg.

Furthermore, and importantly, this was not the sometimes unobserved nucleus or "cytoblast" that Schleiden and Schwann nonetheless believed must be there. Rather, this nucleus was a definite, visible, verifiably differentiated and bounded body within the cell. Additional studies outlined the process: fertilization brings together two cells and two nuclei, one from each parent. Fertilization, in this sense of a combination of two parental contributions, therefore seemed decisive in initiating the development of a new individual life. Perhaps Pope Pius IX was not so far off, even though his proclamation did not follow the biology of the time and even though, as we shall see, our understanding of fertilization and of development has changed since.

The new question was how fertilization works. What is it about the coming together of cell nuclei that makes development and differentiation possible? In the 1890s, Theodor Boveri did not know the answer but was eager to find out. He took a number of sea urchin eggs and embryos after several cell divisions and shook each apart into pieces. His goal was to ask "what happens if . . .?" and then to test a number of different "ifs." What happens such that some of the pieces continue to divide and differentiate, while others just lie still and die? What could he learn about the differences in those that thrived and those that did not? Was there a regular, predictable minimal requirement for sustaining life, and if so what was it?

An outstanding cytologist, Boveri discovered that the only fragments that will develop by growing and differentiating are those with a full complement of chromosomes in the nucleus. He demonstrated this, over and over. Thus, it was not only clear now that mothers carry eggs, that these eggs are cells, and that they are passed on to the next generation, then fertilized to initiate differentiation and the formation of an embryo. It was also now clear that the fertilized egg cell has a nucleus made up of the two parental nuclei that come together during fertilization. Furthermore, that hybrid nucleus must have a full complement of chromosomes. Boveri demonstrated that the material in the nuclei is organized into distinct chromosomes and that each chromosome retains its individuality during cell divisions. This was not at all obvious until he drew on his own and other accumulating studies with definitive demonstrations.[31] Some of the material in the nucleus—specifically in the chromosomes—can be stained; for this reason it was labeled "chromatin." By showing that the chromatin is organized into distinct bodies, or chromosomes, Boveri quickly helped place the presumed locus of heredity right there in those chromosomes.

Boveri's work accorded well with the studies of chromosomes themselves during the 1870s and 1880s. Fol, Hertwig, Walther Flemming, and a growing cadre of other cytologists joined in the exploration of this frontier deep inside the fundamental cellular unit of life. After working out the ways that chromosomes divide during normal cell division, they discovered that the process is quite different when germ cells divide. That is, when an egg or sperm cell divides, its nucleus and the chromosomes in particular go through a much different dance than occurs in dividing body cells. Again, the obvious question was why? As it is today, it was reasonable then to begin with the assumption that if something was complex in a regular and predictable way, there must be a reason. Within science, at least, that reason must be material and natural rather than mystical and supernatural.

In normal dividing body cells (somatic cells), chromosomes double and then separate so that each of the two resulting daughter cells ends up with the same number and same complement of types of chromosomes. This division is mitosis. Reproductive cells (germ cells), however, divide by meiosis; division results in diminution of the material involved. Each reproductive cell ends up with half the number of chromosomes of body cells—not half the number of types of chromosomes, but half the normal number of chromosomes. This works out perfectly when the egg comes together with a sperm that also has a half set. Behold, the fertilized egg has a full set of chromosomes. This beautiful explanation of heredity, combined with development, provided a material, physical, causal, but not vitalistic or teleological account of development.

By the 1890s, it was time for textbook summaries of cell biology. Hertwig in Germany and Edmund Beecher Wilson in the United States published powerful editions that remain astonishing even today in their detailed and careful descriptions and interpretations of the complex processes of cells.[32] Insofar as cells were the fundamental units of living organisms, both structurally and functionally, then cytology was the starting point for all biology. Developmental biology was no exception.

In 1895, just as he was also compiling the first edition of his textbook, *The Cell in Development and Inheritance,* Wilson published his remarkable *Atlas of the Fertilization and Karyokinesis of the Ovum.*[33] This work included ten plates illustrating details of the cytoplasm and of the nucleus and showing exactly what each chromosome was doing through ten selected stages of cell division. Wilson displayed an exquisite scientific technique. What is especially remarkable, however, is that he felt it important to publish these as photographs in what must have been a very expensive project. Collaborating with a photographer colleague at Columbia University, Edward Leaming, Wilson meticulously demonstrated the steps of nuclear and chromosome division that begins the process of cell cleavage. Here was a literal picture of the new theory of heredity and development. By the next year, he no longer felt the need to publish expensive photographs but instead inserted drawings and schematic representations in the second printing.[34]

At about the same time, in 1900, as has been well documented, biologists rediscovered Gregor Mendel's work originally published decades earlier, in the 1860s. Mendelian heredity, as interpreted in the context of 1900, emphasized inherited units that guide development. These units were the material connection across generations that make offspring like their parents. Yet the recombination of genetic units through sexual reproduction also provides opportunities for variation. Thus, we see a balance of conservatism through heredity and innovation through development. By 1910, the Mendelian-chromosome theory of heredity had brought these two lines of thinking together, as we will discuss in the next chapter. Here, finally, it seemed, was a solid basis for modern explorations further and further into the stuff of life and the nature of life itself.

Yet, as Wilson reflected on his earlier ideas during a 1922 lecture in honor of his friend and sometime collaborator William Thompson Sedgwick, "The evidence from every source demonstrates that the cell is a complex organism, a microcosm, a *living system*." Knowing that does not get us very far in understanding life. We look at the cell and see the complex structure of both the nucleus and the protoplasm (or cytoplasm) that surrounds it. This is not homogeneous, undifferentiated matter, but a highly structured, inherited, material cell that includes Golgi bodies, plastids, granules, fibrils, and many more components. "Some of these formed bodies seem to be permanent, others to be transitory formations that come and go in the kaleidoscopic operations of cell-life. Which of them are alive? Which of them, if any, constitute the physical basis of life? What, in other words, is protoplasm?"

"These are embarrassing questions," Wilson admitted gleefully, "for the truth is that the more critically we study the question, the more evident does it become that we cannot single out any one particular component of the cell as the living stuff, *par excellence.*" He went on to show that it is the cells, beginning with the one inherited egg cell, that carry life. As one cell divides into many, the interaction among cells becomes a physical and material living system. The particular interactions and the dynamic processes turn what are fundamentally inorganic particles into the living being. The organization of the whole, in some way, makes up a life.

How, then, do the inherited units in the nucleus work together with the cytoplasm to give rise to an individual life? In their eagerness for an answer, others said that it is "an act of the 'organism as a whole'; it is a 'property of the system as such'; it is 'organization.'" Yet Wilson emphatically insisted that these words mean nothing. Rather, "in the plain speech of everyday life, their meaning is: *We do not know.*" Far from meaning that we cannot discover the answers or that we should turn away from a physical basis for life, however, Wilson insisted that because we do not know does not mean that we cannot know, or that "*ignoramus* does not mean that we must also say *ignorabimus.*" Not now being able to explain how development works in physical and material terms did not mean that it was impossible, or that we should resort to unscientific appeals to a supernatural or metaphysical vitalism. Rather, "Perhaps we should go no further than to record and analyze the existing order of phenomena in living systems, without losing sleep over the imaginary problem of a unifying principle." "For my part, I find it more amusing to look forward to a day when the great riddle may give up its secret."[35] It is to that recording, analyzing, and looking forward that we now turn.

Interpreting Embryos, Understanding Life

By the end of the nineteenth century, each individual life was seen as beginning with an egg inherited from the mother. The egg is a cell. Sperm inherited from the father fertilize the egg, the nuclei come together, and the cell begins dividing into two, then four, then more and more cells, which differentiate and eventually result in a fully formed organism. These cells are the start of life, in the organic sense, but not yet obviously or fully a life. The resulting embryo is neither the epigenetic mixing of fluids that Aristotle had offered, nor the preformed homunculus of Hartsoeker. Instead, the embryo moves the fertilized egg into the fully formed organism and thus holds the secrets of development. The process of embryogenesis (the generation of form throughout embryonic development) is also the process of morphogenesis (the generation of form). By the late 1800s Aristotle's problem of generation of animals had been reinterpreted in material and embryological terms.

If, as Wilson had suggested, the best strategy for biology was to dig in and do a lot of careful observing and analysis, then there was plenty of work to be done. And that work should be focused on embryology. Researchers who could unravel the earliest stages of an individual organism's development might thereby provide the most important clues to how a life begins to be a life and how it becomes life of some particular sort. In 1897, the ninth edition of the *Encyclopaedia Britannica* defined *embryology* as "somewhat vaguely applied to the product of generation of any plant or animal which is in process of formation." The term included all stages, from fertilized eggs to birth, though a human embryo was seen as becoming a fully formed "fetus" at eight weeks, by which time it "has assumed the characteristic form and structure of the parent." Even with the clear goal of studying embryos and embryogenesis, biologists faced the question of how to proceed. What causes development and differentiation, or what explains the emergence of the formed from the unformed? They assessed multiple, sometimes profoundly conflicting ideas concerning what to do next.

One way to study normal development is just that, to study normal development. Go look. This sounds noncontroversial, but as usual it is not. To start with, what should we look at? If we want to study human development but for obvious reasons do not have many samples of normal humans actually developing, do we assume that other animals are just the same? If we have only one sample, when is it legitimate to generalize from a single case, and when can we assume that this case is "normal"? Is it acceptable to assume that each case is similar enough to all others that any one can become the standard for any other, or is it better to assume that individuals vary significantly and consequently that it is necessary to compile multiple samples and to standardize across variations? In other words, how can we discover the "normal"; how can we be sure that any one sample is nearly enough normal that it can serve as the norm? Epistemological assumptions and interpretations necessarily abound.

WHOSE VIEW OF LIFE?

Methodological differences, and differences in why researchers wanted the data, also colored the choice of research approach and the results. Three approaches dominated. First came the comparative morphologists. These men (and they were men) collected embryos from a diversity of different species and sought to describe and collate the developmental stages of each before comparing the results. What do we learn from this comparison, they asked? Until the mid-nineteenth century, they saw evidence of design and regularity in nature.

Again, nineteenth-century embryologist Karl Ernst von Baer gave us a framework for thinking about embryonic development in the context of species development. He was responding to those who, motivated by their philosophical convictions to find unities and general laws in nature, had recently suggested that each individual developing embryo follows a sort of recapitulation of simpler forms. Superficially, this idea made sense as a way to find generalities amid apparent complexity and diversity. If, indeed, every type of organism followed a pattern of development found also in other types, this similarity would demonstrate a tidy unity of nature. It would suggest that the simpler forms might provide clues to understanding development in the more complex and "higher" forms, as they were then considered. An intelligent designer might well have used an effective set of patterns and processes to guide the early stages of development in every type of organism. Von Baer strongly disagreed.

In his marvelous *Über Entwickelungsgeschichte der Thiere* in 1828, von Baer offered five "scholia," designated as laws by his followers and summarized as establishing:

1. That the more general characters of a large group of animals appear earlier in their embryos than the more special characters.

2. From the most general forms the less general are developed, and so on, until finally the most special arises.
3. Every embryo of a given animal form, instead of passing through the other forms, rather becomes separated from them.
4. Fundamentally, therefore, the embryo of a higher form never resembles any other form, but only its embryo.

And, finally, "The history of the development of the individual is the history of its increasing individuality in all respects."[1] Von Baer refuted the notion that higher organisms copied the developmental stages of lower organisms, as those seeking unity insisted they did. Instead, von Baer saw divergence and differentiation. His persuasive and persistent articulation of this view, drawing on considerable evidence from his own observations, became important in later debates about evolution.

In 1859, with his theory of evolution of species by common descent from vastly distant ancestors, Charles Darwin provoked controversy, of course. Most readers of the *Origin of Species* focused on assessing its implications for man's place in nature. Darwin himself had seen embryos as particularly illuminating for that and other fundamental questions about life. For Darwin, persistence of rudimentary organs like the appendix in humans, aspects of the fossil record, and parallel (or morphologically homologous) structures in different species all provided evidence in support of his evolutionary interpretation.

Darwin held embryology in the highest regard. In Chapter 13 of the *Origin* he asked, in typical Victorian prose: "How, then, can we explain these several facts in embryology, namely the very general, but not universal difference in structure between the embryo and the adult; of parts in the same individual embryo, which ultimately become very unlike and serve for diverse

WHOSE VIEW OF LIFE?

purposes, being at this early period of growth alike; of embryos of different species within the same class, generally, but not universally, resembling each other; of the structure of the embryo not being closely related to its conditions of existence, except when the embryo becomes at any period of life active and has to provide for itself; of the embryo apparently having sometimes a higher organisation than the mature animal, into which it is developed." In fact, he concluded, "I believe that all these facts can be explained, as follows, on the view of descent with modification." Furthermore, "the leading facts in embryology, which are second in importance to none in natural history, are explained on the principle of slight modifications not appearing, in the many descendants from some one ancient progenitor, at a very early period in the life of each, though perhaps caused at the earliest, and being inherited at a corresponding not early period. Embryology rises greatly in interest, when we thus look at the embryo as a picture, more or less obscured, of the common parent-form of each great class of animals."

By revealing parallels not only in adult structure but also in the developmental stages and in the processes of development, embryology considerably undercut the theory that each species is created separately and for its own designed purpose. Instead, the embryological evidence of consistent and significant parallels argues for common descent. If all species descended from one, we would expect the parallels we see. If every species is created separately, we would not. Therefore, the new theories of evolution with common descent and embryology fit together, both having as a theme the unfolding of forms.

In Germany, naturalist Ernst Haeckel took up this argument. He offered a view of life, including man, based in his philosophy of "monistic materialism." Always one to take a broad sweep rather than to suffer through hours in the laboratory observing details, Haeckel sought generalizations that would reinforce evolution. For our purposes, the most important result was his biogenetic law. "Ontogeny is the brief and rapid recapitulation of phylogeny," Haeckel asserted. That is, each individual in the course of its own development follows and even literally passes through developmental stages of the species in its evolutionary history. Rather than a perfect recapitulation, which would obviously take a long time, individual development takes place with some abbreviated steps.

This idea, articulated initially and most forcefully in his *Generelle Morphologie,* held great attraction for those eager to have not only a description but also an explanation for development.[2] Recall that in the 1870s, 1880s, and 1890s, cytologists were busily studying the cell and were just discovering the role of the nucleus, the structural complexities of the cytoplasm, and other details that soon undercut what came to be seen as Haeckel's wild generalizations. But they were only beginning this study, and they were busy inside the labs, while Haeckel was out in the field and, most importantly, in the public. When Haeckel first articulated his ideas, his impressive theoretical system seemed seductively neat and useful to many. It provided a richer and more provocative framework for public discussion and debate than esoteric laboratory descriptions could possibly offer.

The picture Haeckel presented was literally that: a picture. His readers could see laid out before them neat embryonic sequences of several vertebrates, including man (see Figure 4). They looked nearly enough alike to the lay reader, and differences in detail just seemed to show that those were mere details. The proposed explanation held considerable appeal. Of course we should expect these embryos to look alike and to pass through basically the same stages in development: they had a common cause. That cause was the historical continuity brought by evolution. There

WHOSE VIEW OF LIFE?

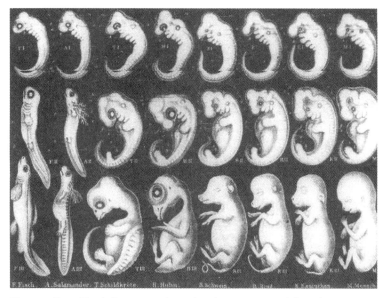

Figure 4. Ernst Haeckel's comparative embryology. From Haeckel, *Anthropogenie oder Entwicklungsgeschichte des Menschen* (Leipzig: Englemann, 1894).

is considerable debate about the extent to which Haeckel either did see or could have seen what he claimed, and about the extent to which he may have faked his data to construct these series of parallel embryonic stages. If he did falsify data, there is dispute about whether he did so purposefully and willfully or whether he was simply so persuaded by his theory that he really saw things that way. For our purposes, what matters was his impact, which was probably less hurt than enhanced by the controversy—at least among the public.[3]

Haeckel went further than just pointing to parallels. Let's trace back development to the earliest significant embryonic stages, he said, or what was, in effect, the beginning of the individual life. What we find is a series of early stages, and one that is the most primitive of all: the gastrula. All animal embryos start as a gastrula, Haeckel argued, which is then divided into "germ layers." Haeckel was convinced that the gastrula was the first significant developmental stage. Before that, the egg was just matter, Haeckel insisted; it is the germ layers that begin to define the embryo and bring it to life. This notion was consistent with the traditional view that an individual life really begins at forty days or so. Haeckel pointed to the gastrula and following stages instead and suggested that in the first month there was just material, vegetative change involving growth.

Haeckel even postulated the existence of a first organism in evolution: the gastrea. His gastrea theory seems ridiculously simplistic today, as it did even at the time. Nonetheless, the desire to discover a material starting point for each individual life was shared by many. Haeckel gave the embryo what he saw as a material motive cause through evolution, since he believed that the force of history was, in effect, packed into that gastrea. It unfolded according to its internal material direction. Then evolution provided an environment that called for adaptations, and each step of evolution added those adaptations to the end of development. The earlier stages remain essentially the same, though they might proceed more rapidly in early development than might be expected from the species' history. Only the end result of development displays significant changes. Haeckel's was an enchanting theory in many ways. It generated much lively discussion, eager study of embryos, wide public attention in Germany and the United States, and controversy from many directions.

Critics like von Baer rejected Haeckel's interpretation of recapitulation and his emphasis on parallels in embryology. Von Baer was certain that his own careful empirical study showed without doubt that the early stages vary considerably and are *not* parallel. He saw Haeckel's as just a newer version of the old recapitulation theory that von Baer had been combating decades earlier. Furthermore, von Baer rejected Haeckel's materialistic evolutionary interpretation in favor of creation. For von Baer, evolution certainly provided no easy answers to the causes of embryogenesis.

Other careful observers challenged Haeckel's insistence that the gastrula is the earliest developmental stage that matters. They asked why not go back to the beginning—the beginning of the individual organism, that is. Why begin only with the gastrula and germ layers, they asked, when it was reasonable to assume that the earliest cellular structure and the cell cleavages might well also hold significance for embryo development. Why not start with the single-celled fertilized egg? This did not mean that form or a life began at that point, but rather that the material basis for life began with the joining of sperm and egg. Whether preformation or epigenesis would prevail remained an open question.

What emerged was a group of meticulous researchers staring through microscopes for hours and hours, day after day during breeding seasons, when they could study the embryos of many organisms at the same time. Some investigators traced cell lineages, asking exactly what happens at every step of the way during development. What happens when the egg cell divides into two, or into four? What patterns of cleavage, rotation, or other details can we see?

The leading cell lineage researchers were Americans, most of whom gathered at the Marine Biological Laboratory at Woods Hole, Massachusetts, during summers in the 1890s. Charles Otis Whitman, the director, led his students and junior colleagues to the comparative study of embryos. The embryos of marine invertebrates were favorites, as they were readily available, develop outside the body, and are easy to observe. The idea was that if enough individual researchers focused on enough different types of organisms, the collective comparative data would reveal the patterns of development and the extent of similarities and dissimilarities.

This sounded like a great idea, but it demanded an astonishing amount of meticulous hard work. Furthermore, such study required a number of background assumptions about what is actually being seen through that microscope. For some organisms, the egg is large and transparent. Though not ideal, it is possible to set up a microscope with natural light and sketch the changes that occur on the surfaces and even within cells of this type. Other eggs are not so transparent, and only the external surface remains visible. Documenting the internal details involves killing, fixing, preserving, and then examining a lot of eggs. This is painstaking work, since observing each cell division requires a different preparation.[4] Furthermore, eggs do not wait for a convenient time of day to begin developing. They start when they are fertilized. Many biologists' spouses discovered that summering in Woods Hole meant either visiting at all hours in the lab, if they wanted to see their loved ones, or waiting at home alone. Wilson's wife noted that though she found this invasion into their newly married private lives "very distressing," it was understood and accepted as part of the job.[5] Couples got used to the demands, or the relationships did not last.

The result of all this work was instructive. Far from a sequence of neat parallels, embryological development displays many differences. Eggs from different types of organisms divide differently, and even individual embryos within the same species might behave differently in certain particulars. Some divisions are neat and regular, others spiral; some result in cells of the same size, some much smaller, and others much larger cells. Considerable diversity and complexity prevails in the natural

WHOSE VIEW OF LIFE?

world. Yet it was clear that even the earliest cell divisions were significant for later development. Even if the egg was not already formed and a life really began later, the first cleavage could influence the patterns and the details of the developmental steps that followed. What was clear in all this flood of investigation and resulting data was that Haeckel's simplistic generalizations were seriously flawed—not merely too simple, but wrong. It is not the case that every individual form originates only at the gastrular stage, driven by evolution. The earlier cell divisions matter too. Because these early divisions are not all parallel in different species, the cell lineage group rejected evolutionary recapitulation as a causal explanation of individual embryogenesis.[6]

This was exciting and important biological work, and it laid the foundation for developmental biology today. Yet, in the end, all the descriptions produced in these years did not take embryology very far, nor contribute directly to a consensus on when a life begins. It was simply too difficult to interpret the results. While it was clear that phylogeny does not materially cause ontogeny in any direct or useful way, it was still not clear what does. This uncertainty left considerable room for more debate.

By the early twentieth century, then, biologists had modified both preformationism (now seen in terms of predetermined inherited material passed on from parents to offspring through inherited particles called "genes") and epigenesis (the material unfolding of preorganized "organ-forming germ-regions" or the expression of relevant "fates" by cells and germ layers). Determining the balance of predirection and epigenetic emergence was, as Oscar Hertwig put it, *The Biological Problem of Today*.[7] As biologists prodded, sliced, and diced developing organisms from the germ cell stage through fertilization, they began to identify what looked like stable stages, which they named the "morula," "blastula," "gastrula," and so on. Embryogenesis might well be a regular and predictable process, even if we could not see or explain every step of the way. It also seemed relatively unimportant what was called an "embryo" or what the precise domain of embryology was thought to be. When did a fertilized egg become an embryo and when did an embryo become a fetus, or what was the definition of the embryo? It was not clear that it mattered much, and concern about embryogenesis gave way to thinking about emergence of form, or morphogenesis more generally. The "embryo" faded in importance, as a sort of placeholder, while the series of defined stages and the patterns and processes of change gained prominence.

Picturing Embryogenesis

The resulting approach to development focused on identifying and standardizing stages of development. Establishing regularities fulfills a medical purpose, since if we can hold up a chart of what is supposed to be normal, then we can tell whether a particular example deviates from the norm. That is potentially extremely useful. It is also useful to be able to compare across species. If other animals are basically like us, then we can learn a lot about the subtleties and inner workings of development by cutting open those other animals. Since we obviously can't go around cutting open humans, animal studies are a valuable tool.

Cell lineages served as a set of normal tables, but only for the very earliest stages of development and only for a few organisms, not including vertebrates with internal eggs. They were of limited value since it still was not clear what significance those earliest developmental steps had for later development of form. Thus, we find frustrating efforts to generate descriptions of the normal sequence of events, through a wider sequence of stages. These began with assumptions that there are, in fact, normal stages and a normal sequence and that we can discover them. The representations of results included drawings, photographs, preserved specimens, and three-dimensional models, in an attempt to make the findings accessible to both biologists and medical practitioners.

Wilhelm His made the first major contributions to our understanding of the details of human embryology. The first plate in his *Anatomie menschlicher Embryonen* in 1880 pictured twenty-five stages of human development.[8] It is easy to imagine this figure as a sequence of snapshots of the same embryo, and that was presumably the intention. Of course, each stage necessarily comes from a different embryo, since once the embryo is killed and preserved it cannot continue growing. These stages cover the first two months of life, just up to the point where the embryo is labeled a "fetus," because it now finally has all the human parts and the beginning of a recognizably human shape (see Figure 5). As historian of science Nick Hopwood has explained, "His created human embryology as it would be practiced through the twentieth century."[9] Hopwood has also studied the two series of wax models and ten glass photographs sold separately to accompany the textbook. Through his marvelous representations and descriptions, His created and continued to dominate embryology as a productive science. His's techniques and methodological practices shaped the discipline of modern human embryology.

In particular, His's sectioning techniques and microtome made it possible to see inside human embryos much better than anyone had been able to do before. Since so few embryos were available, His compiled all the samples he could find. His approach involved destroying the whole embryo in favor of the slices, so he had to persuade others that doing so was a sensible expenditure of limited resources. This could not have been obvious to

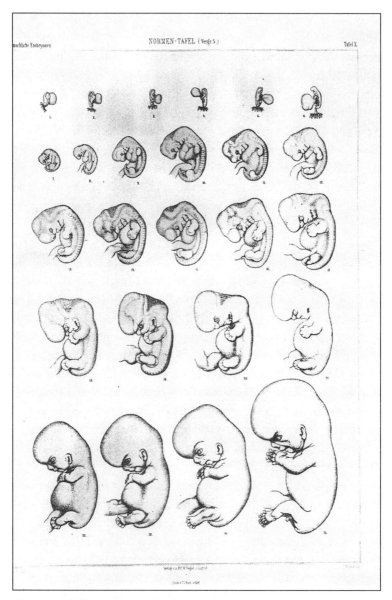

Figure 5. Wilhelm His's human developmental stages. From His, *Anatomie menschlicher Embryonen,* vol. 3 (Leipzig: Vogel, 1885).

everyone. Indeed, as Hopwood explains, His had to enlist help from many different people and to create a network of suppliers of embryos.

Specimens could come from abortions—mostly natural and spontaneous but sometimes induced—or miscarriages in which either the mother herself, a physician, or a midwife would have to collect and preserve the embryo. Not surprisingly, nearly all embryos came from physicians, as mothers were not likely to be thinking of such things and midwives were less likely to be connected to the collection network or to consider the collection important. Without sufficient reward, there was little incentive to collect. Some specimens came from postmortems of pregnant women, again obviously through the physicians who did the autopsies.

For His to collect these samples, he obviously had to enlist physicians' cooperation. To be successful, His had to situate himself at what Hopwood calls "the center of a supply network of scientists and physicians." His's argument, one that yielded an impressive collection of seventy-nine embryonic "treasures," was based on a different approach than had been taken before. Again, as Hopwood explains: "More innovative than the scale of the operation was the way he linked means of preserving and transporting specimens, and the new methods of analysis for which he was campaigning with a moral duty to give material to him." This was a remarkable claim on His's part: that physicians had a moral obligation to contribute to his research. Making it required considerable self-confidence or even arrogance on his part. But he was evidently persuasive and his approach worked. In his introduction, His "argued that these exceptionally valuable specimens must be subjected to destructive analysis with state-of-the-art techniques. Gynecologists, [who] he claimed had been wasting or ruining the 'precious objects' to which they had privileged access, should send them to him."[10] But why should they do it? Because His agreed to name the embryos, or what he called the *Menschlein,* or "little men," after the donor. Not, as Hopwood points out, after the mother, but after the physician.

Once he had his precious treasures, His made sure that they were adequately preserved and fixed so that they would remain stable and survive over time. He then sliced them with the supersharp blades in his microtome to produce a series of regular-sized cross sections. He then could reconstruct the embryo to produce the whole. This was a new whole, one that was only virtually observable and constructed out of pieces, but nonetheless a whole. The technique required considerable skill and numerous assumptions, but others followed His down this path and agreed that the assumptions and the hard work were reasonable investments.

Improved microscopes, paraffin for preserving specimens to get cleaner and thinner slices, and advanced methods of reconstruction all helped to enlist a cadre of anatomists interested in following His into human embryology. While there were controversies about interpretation, about collecting the embryos, and about their significance, His had at least two important students who carried on the work of establishing normal human stages even further.

Setting out the normal stages was just the start, of course. No series could be perfectly complete, so collection and description must continue. And since each specimen has some variations, it was necessary to collect more and more examples to "normalize" or standardize the sequence of stages. Further, after describing them, it was time to study the stages in detail in order to interpret both their structure and their function. His's student Franz Keibel took up all these objectives.

As Keibel noted, His had intended to work out a developmental account of the complete human body, but "as time went on the hope of accomplishing the task single-handed failed him; and so he suggested that I should collaborate with him in writing a text-book on human embryology." Even that was too much. In collaboration with Curt Elze, Keibel did manage to produce and publish a series of normal stages in human development, from about the twelfth day through the second month. This is a crucial period, from implantation in the uterus to the formation, at least rudimentary, of all the major body parts in the fetal stage.[11] Though not the earliest stages, this series represented a much earlier sequence in much more detail than had been available before.

Like His, Keibel noted the cooperation of a network of researchers. As he put it, "I had very striking evidence of the community of purpose which to-day inspires our scientific world. A considerable number of investigators deprived themselves for long periods of time of valuable material which they themselves had not yet had opportunity to study thoroughly, in order that a larger undertaking might be completed."[12]

By 1910, Keibel discussed how much further the possibilities for progress had come, for "to-day the post and the telegraph, railways and steamboats, are at our service and may be turned to the service of science." These new tools made it possible for Keibel in Germany to work with His's leading American student, Franklin Paine Mall. Together, they produced their remarkable *Manual of Human Embryology,* and as professor at the Johns Hopkins University from 1887 to 1910, Mall also developed a collection of embryos housed at the Carnegie Institution of Washington. This collection eventually grew to over 10,000 embryos, now at the National Museum of Health and Medicine of the Armed Forces Institute of Pathology, and has recently been extended with the aid of modern imaging techniques as the "Visible Embryo Project." In 1987, dedicating their work to the memory of His and Mall, Ronan O'Rahilly and Fabiola Müller summarized the earlier efforts at establishing normal stages.[13]

In the early 1900s, deciding what was meant by an "embryo" or "organism" became less important than determining to what extent the stages were defined, predictable, and determined rather than regulated in response to environmental conditions. Textbooks of human embryology outlined successive stages, focusing on the appearance of organs and visible features such as limbs and face. Both because of what was possible with existing research approaches and because of what medicine deemed important, later fetal stages remained the primary focus.

Keibel and Mall admitted that they knew little of the early stages: "Nothing is known concerning the fertilization of the human ovum, but it may be presumed that it takes place in essentially the same manner as in other animals." "The segmentation stages of the human ovum have not yet been observed." And so on. They were restricted to what they could see with the few selected specimens they had, starting with the fourteen-day "ovum," which itself revealed little. In later stages they could focus on recognizable organs and the sequence in which they arise. Yet even here, the authors noted,

> It is clear that the normal fully developed organism cannot be produced without a certain regular succession in the development of the organs and a regular interdependence of the individual developmental processes, but it is open to question whether this interdependence is the result of the individual and independent development of each organ taking place in such a way that it fits into that

WHOSE VIEW OF LIFE?

of the others or whether it is due to the individual anlagen [structural rudiments] of an organism mutually influencing one another during the development so as to cause the formation of a normal organism.[14]

These leading human embryologists used the terms *ovum* and *organism* while focusing on the succession of parts and organ. They largely yielded the term *embryo* to nonscientific usage. Biologists instead considered "morphogenesis," "organogenesis," "growth," "differentiation," and such processes. By the 1960s, "embryology" had largely disappeared from biology, along with explicit discussion of embryos; the field had given way to "developmental biology." Encyclopedia definitions also became looser. For example, the current *Britannica* defines *embryo* as "the early developmental stage of an animal while it is in the egg or within the uterus of the mother. In humans the term is applied to the unborn child until the end of the seventh week following conception; from the eighth week the unborn child is called a fetus." All the work that had been done on normal stages was only a beginning. Many questions remained about how to interpret the developmental stages and what caused the development over time.

Experimenting to Reveal the Normal

Finally, drawing on all the above perspectives, comparisons between the pathological or experimental and the normal led to another approach to "seeing" and interpreting normal development. Where it was not possible simply to see everything directly, the researcher could contrive experimental conditions to create new access to normally hidden secrets and ask, "What will happen if I do such-and-such manipulation . . . ?" The field of experimental embryology arose rapidly at the end of the nineteenth century as a result of such opportunities. While experiments cutting open humans could not be appropriate, of course, experimental study of other animals was seen as providing valuable information regarding the otherwise largely inaccessible human development.

Experimental embryology really began with the work of Wilhelm His, Eduard Pflüger, Gustav Born, and others who wanted to address questions that they could not answer through direct observation. It gained a name and visibility with the enthusiastic advocacy of Wilhelm Roux. As often happens, the field crystallized around a particular event, person, or publication. In 1895, Roux started a journal, which he not so modestly entitled *Roux's Archiv für Entwicklungsmechanik der Organismen*. This became the journal of record for several years, until many embryologists became frustrated with Roux's heavy-handed editorial practices and started other publications, including the American *Journal of Experimental Zoology* in 1903. Having a journal with a special word, *Entwicklungsmechanik* (developmental mechanics), gave credence to the new field and attracted younger scholars to take up the experimental and analytical study of embryology.[15] While not the first to do experimentation in embryology, Wilhelm Roux led the way in explicitly calling for an experimental research program.

One pivotal experimental episode, which still appears in textbooks and class discussions today, demonstrates the power and limitations of experiments. Roux and another German embryologist sympathetic to Roux's approach, Hans Driesch, each experimented on the two-cell stage of an organism: Roux with frogs and Driesch with sea urchins. They asked the question: what will happen to one of the cells if it is separated from the

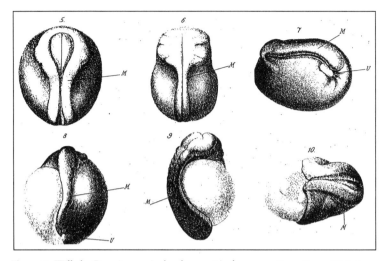

Figure 6. Wilhelm Roux's mosaic development in frog eggs. From Roux, "Beiträge zur Entwicklungsmechanik des Embryo," *Virchows Archiv für Pathologische Anatomie und Physiologie und klinische Medizin* 114 (1888): 113–153.

other after the first cell division? Roux used frogs because they were readily available and had large, visible eggs. With frogs, however, the eggs have a relatively solid membrane that holds the cells together through a number of cell divisions (see Figure 6). This makes sense, since it keeps the separate cells from flying apart if they are knocked about in a pond. The membrane made it difficult to separate the two cells after the first division, so Roux did the next best thing he could imagine. He stabbed one of the two cells with a hot needle to kill it and therefore, he believed, to allow the other cell to develop independently. He did this, he noted, to "solve the problems of self-differentiation—to determine whether, and if so how far, the fertilized egg is able to develop independently as a whole and in its individual parts. Or whether, on the contrary, normal development can take place only through direct formative influences of the environment on the fertilized egg or through the differentiating interactions of the parts of the egg separated from one another by cleavage."[16]

Roux discovered that the remaining cell develops essentially as it would if it were part of the normal embryo to the gastrula stage at least. It is, he interpreted, in effect a half-embryo. This reinforced his conviction that development occurs in a mosaic-like manner, with individual cells adopting individual fates in the whole organism that follows, just as each mosaic tile has its independent characteristics but has a particular role in the resulting pattern. For Roux, this also meant that the causes of development lie inside the cells, and inside the egg from the beginning—somehow—rather than coming from outside or environmental influences. Development and differentiation were much more a matter of internal material factors and inherited determinism than responses to environmental changes. The normal process, he concluded, is predictable, and therefore an experimental program that could demonstrate how each part works would reveal the causes of development.

Intending to follow up on Roux's work with a different organism, Driesch took the two-celled stage of his sea urchins and shook apart the two cells (blastomeres). This seemed a much better experiment to accomplish what Roux had wanted. From earlier studies of sea urchins at the lovely seaside research laboratory in Naples, Driesch knew that their membrane is much more fragile. It is possible, by shaking sea urchin eggs after early cell divisions, to separate the cells. He did that and watched. As he noted later, he looked forward to seeing half-embryos. He expected to find the same results that Roux had. Instead, each of two and then four cells grew into a complete, even if a little smaller than normal, larval form. They did not at all act as if they were just half of a mosaic, but rather as if each were a reor-

WHOSE VIEW OF LIFE?

ganized whole.[17] Driesch concluded that each blastomere at the two- and even four-cell stage is what he called "totipotent," or capable of developing into a normal whole despite changing condition. That is, it demonstrated considerable ability to "regenerate" the whole in response to changing conditions both inside and outside the organism.

These conflicting results raised fundamental questions. Though Roux felt comfortable adding various hypotheses about "postgeneration" and "auxiliary" sets of inherited determinants, others rejected what they saw as ad hoc inventions. They wanted to address the core questions about what the process of differentiation really involves and what causes it. Are cells already set off in different directions, or differentiated, at early stages of division? Or are they capable of retaining their flexibility until later? Are they normally differentiated but nonetheless capable of reprogramming later, in response to changing conditions? And if so, for how long do they remain undetermined or "programmable"?

These are the same questions we are asking today, over a century later, about stem cells and cloning. Just as today, in the 1890s the excitement of apparently conflicting results raised fundamental questions about the nature and significance of differentiation. A rush of activity followed, as biologists first took up parallel studies in other organisms for comparison and then new studies to get at the same questions or to use the experimental approach in other ways.

Roux or Driesch? Mosaic or regulation? Inheritance or development? Frogs or sea urchins? Which was dominant; which was right? Instead of arriving at neat answers to settle the questions, together Roux and Driesch discovered a diversity of developmental patterns. Some organisms followed Roux's pattern and differentiated early, in a more mosaic-like way. Others were more like Driesch's sea urchins, more flexible and able to respond to changes in the surrounding conditions while still developing normally. Hans Spemann showed that in amphibians, for example, if the researcher constricted the fertilized egg in one plane using a very fine loop, he could produce identical twins side by side in the same egg. If he constricted the egg at the plane ninety degrees to the first, he got one cell with a developing embryo and another full of dead junk. Still another set of partial constrictions produced imperfect twins, joined where they had not been fully separated.[18] News like this suggested a mixture of determinism and regulation. Now there was a new set of questions: what determined which pattern a particular organism follows? Would a sea urchin always behave as Driesch's urchins did, or were there some conditions under which they would be more mosaic-like, for example? If an organism can regulate itself, does that mean that there is a self to regulate, and is this perhaps what we should mean by a life? Furthermore, how could we know?

Experimentation clearly held the most important key at the time. In the period between 1870 and the 1890s, researchers had already made tremendous advances in understanding patterns of the earliest stages of normal cell division and differentiation through cell lineage studies. Yet they soon reached the limits of that research program and had to set aside what they recognized were exciting questions about what causes the patterns and what processes guide them. Experimental manipulations and methodologies provided the framework in which they could gain additional knowledge. It is worth looking at some of these approaches in detail, focusing on regeneration, parthenogenesis, transplantation, and explantation (transplantation to a culture medium outside the body).

Regeneration

Regeneration provided a sort of natural experiment. That is, some organisms are capable of repairing or regrowing their bodies after being chopped up or losing body parts. Earthworms and planarians, both readily available and easy to study, served as favorite research subjects. In January 1900, Thomas Hunt Morgan began a series of five lectures on regeneration at Columbia University. Better known for his later Nobel Prize–winning research on the genetics of the fruitfly *Drosophila,* in 1901 Morgan summarized the existing literature on regeneration in his textbook on the subject. There, Morgan very decidedly rejected Roux's mosaic interpretation of development. At this point in his career, Morgan was a committed epigeneticist, and Roux's interpretations smacked of an unwarranted preformationism.

August Weismann's important study of the *Germ Plasm* accorded with Roux's interpretations. What came to be called the Weismann-Roux (or, according to Roux, the Roux-Weismann) theory of development began with inherited germ plasm. This germ plasm, residing in the egg cell's nucleus and then augmented by the sperm's nucleus, consisted of chromosomes. Chromosomes, Weismann and Roux agreed, carried inherited determinants, each of which contributed to determining a character in the developing organism. It is easy to read too much into this theory and to interpret these determinants as what we know as genes. They were not genes however; Weismann's complex theory involved a hierarchy of different inherited material particles that competed with each other in a natural selection–driven struggle among the parts, or a "Kampf der Theile."[19] According to this view, those particles inherited in the germ plasm competed among each other and then directed through the process of development and differentiation to determine which details an individual organism will have: what patterns on the wing of a butterfly, for example, or what color something is.

Weismann devoted a section of his book to regeneration, arguing that the process occurs in some organisms, like earthworms or crabs, because of evolutionary adaptations to the struggle for existence. These organisms often experience damage that, unless repaired, would be disastrous. As a result, over time they have developed the capacity to regenerate the important parts that they are susceptible to losing. This happens, Weismann suggested and Roux concurred, because of the "reserve idioplasm," or a backup set of determinants for these parts. Everything could be explained in terms of the inherited particles in the germ plasm. For Morgan, their theory was both an unsubstantiated hypothesis and preformationist. He found the interpretation unjustified on both grounds.

On appropriate scientific practice, Morgan wrote that "when we leave the analytical method and attempt to construct injudicious theories that make the pretense of explaining a complicated process without attempting to resolve the process itself into factors, then progress stops. Such, I believe, to be the case in the attempt to explain the process of regeneration by a theory of preformed imaginary germs. Any theory of this kind is only a pretense; imagination takes the place of verifiable hypothesis, and the process that we set out to study is explained by saying that there are 'germs' present that have been set aside to bring about the result!"[20]

Instead of such unwarranted speculation, Morgan called for an explanation in terms of simple physical or chemical interactions. The "internal relations" of the parts within the whole organism work to regulate and redirect the parts, as needed. Morgan's interpretation was very much closer to that of his friend Driesch, in that it allowed for considerable regulation in response to changing environmental conditions and rejected the possibility of an unyielding deterministic heredity. As Morgan saw it, development was not directed by inherited particles arranged along chromosomes but instead occurred through physical and chemical reactions. Heredity is not just a matter of transmitting material units that then determine what follows, but rather a series of reactions and interactions of the whole organism.[21]

It is important to realize that this interpretation made good sense and was good science at the end of the nineteenth century and beginning of the twentieth. The fact that even this founder of modern genetics saw the important role of regulation and of gradual epigenetic differentiation rather than inherited preformed and preprogrammed development tells us much about the time. It also serves as a useful warning that we not become complacent today in thinking that we have all the answers about development or any other scientific problem. Progress occurs, things change, we learn more, and sometimes what we learn challenges our old cherished views. Often, familiar questions remain even while they take on new meanings or new explanations. Discussions about stem cells today take us back to the same root questions that Morgan was asking, since for stem cells to become whole organisms a sort of regenerative process must take place.

Parthenogenesis

At the end of the nineteenth century, parthenogenesis (development without fertilization) offered another challenge to established interpretations of how life works, just as new forms of parthenogenesis are doing today. Recently arrived from Germany and relocating his summer studies from the Naples Zoological Station to the Marine Biological Laboratory in Woods Hole, Jacques Loeb took up the challenge. He was very interested, as Morgan and Driesch were, in what role external conditions play in development. To what extent and in what way do the surrounding conditions influence what happens within embryonic cells? Since Loeb knew of research by the botanist Julius Sachs that showed that osmotic pressure in the water made a difference for plant growth and differentiation, he resolved to explore whether osmotic pressure might play a similar role for animals. In Woods Hole he tried a number of experiments, including one in which he placed sea urchin eggs into a variety of different saltwater concentrations. Focusing on the earliest cell divisions and developmental stages, he asked what would happen in the different conditions.

What he found when he placed fertilized eggs in a highly salted solution and then back into the normal seawater concentration was instructive. While the eggs were in the salted water, they appeared to rest. When they were returned to the normal water, however, they divided into a large number of blastomeres very rapidly. They reached the blastula, gastrula, and even pluteus larval stages. This suggested to Loeb that during the resting period, the nuclei had been dividing and preparing for cell division and even differentiation. Furthermore, by 1899 he showed that under some conditions even unfertilized eggs could begin to divide and proceed through a number of cell divisions up to the larval stage, though at a speed different from that of normal eggs.[22]

Artificial parthenogenesis: a laboratory scientist uses simple experimental methods to cause an egg to develop. No fertiliza-

WHOSE VIEW OF LIFE?

Figure 7. Reception of Jacques Loeb's parthenogenetic development from unfertilized sea urchin eggs. *Chicago Sunday Tribune,* Nov. 19, 1899, p. 33.

tion needed. No male nucleus needed. This was startling, indeed, and the popular press loved it. "Creation of Life. Startling Discovery of Prof. Loeb's Lower Animals Produced by Chemical Means. Process May Apply to Human Species. Immaculate Conception Explained. Wonderful Experiments Conducted at Woods Hole," declared the *Boston Herald.* The *Chicago Sunday Tribune* announced: "Science Nears the Secret of Life."

The secret of life. Had Loeb discovered this secret? Did he even think he had? Loeb himself asked "What conclusions may we draw from these results? If we wish to avoid wild and sterile speculations"—which of course Loeb did, since he felt that it was important to restrict interpretations and stick to the facts— we must stick with answerable questions. Perhaps heredity and development of a formed life had causes other than the stimulation of fertilization and cell division. We cannot know more without detailed experimental research, he urged, and so "I consider the chief value of the experiments on artificial parthenogenesis to be the fact that they transfer the problem of fertilization from the realm of morphology into the realm of physical chemistry."[23]

For Loeb, "the problem of the activation of the egg is for the most part reduced to physico-chemical terms." The egg sits there, ready to be activated by a single sperm. If, however, no spermatozoon enters the egg, it will die, and fairly soon at that. Fertilization, in effect, prevents the egg's death. It does this specifically through chemical action. For Loeb, chemistry provided the clue to the "riddle of life." He rejected older ideas that some "life principle" enters the body as musings of primitive and pre-scientific man. "Scientifically, however, individual life begins (in the case of sea-urchins and possibly in general) with the acceleration of the rate of oxidation in the egg, and this acceleration begins after the destruction of its cortical layer. Life of warm-blooded animals—man included—ends with the cessation of oxidation in the body." Life begins and ends with the cycle of oxidation. Therefore, "the problem of the beginning and end of individual life is physico-chemically clear."

In hindsight, Loeb's interpretation was extreme and his emphasis on oxidation as *the* key to life unwarranted, though at the time he offered a coherent account. He saw every aspect of life as falling within a mechanistic framework. Therefore, even the "wishes and hopes, efforts and struggles, and unfortunately also the disappointments and suffering" that make up "the contents of life from the cradle to the bier" were results of oxidation and mechanics. Why should not "this inner life" be physically and chemically analyzable, he asked, since researchers had already gone far toward explaining basic instincts that way. Why not also hunger, sexual desire "with its poetry and its chain of consequences, the maternal instincts with the felicity and the suffering caused by them, the instinct of workmanship," and those other

WHOSE VIEW OF LIFE?

instincts "from which our inner life develops"? Though he did not yet have the answers, he was convinced that his was the right approach and that it was just a matter of time before scientists had mechanical explanations for all life phenomena. Even our human sense of proper behavior, as exhibited in ethics, would derive from physical and chemical causes and be mechanically explained. Indeed, "Not only is the mechanistic conception of life compatible with ethics; it seems the only conception of life which can lead to an understanding of the source of ethics."[24]

Historian Philip Pauly explores such questions in his marvelous study of Loeb and concludes that Loeb was very much intending to produce artificial parthenogenesis.[25] This was not an accident, as he sometimes claimed later, and it demonstrates Loeb's conviction that life is material, mechanical, and able to be engineered. Even the most apparently subtle processes of conception and fertilization are not the start of life on this view. Life is the material processes that carry the inherited egg on to become a formed and differentiated organism, shaped in response to changing environmental conditions and tractable to human manipulation. For Loeb, apparently, these processes *were* life. Yet the fact that a scientist in the laboratory could change some chemicals, do some manipulations, and cause at least some of the processes that are considered to constitute life called into question what we mean by a life. Conventional definitions needed more fine-tuning in the face of artificial parthenogenesis, perhaps.

That scientists were not only engaged in the study of life by watching, describing, and recording details but also thinking of ways to "engineer" better life: Pauly makes the persuasive case that this was Loeb's view. Others felt the same way, though few working on human biology went as far as he did. Glimmerings of the possibilities for human intervention in natural processes were beginning to shine at the dawn of the twentieth century, nonetheless, just as they seem to be doing today. But at that time, the dark side of manipulation was less clear. The progressivist movement that brought hopes of progress for all through science and technology held the day, among those who were paying attention to such questions at all.

Transplantation

Back in the laboratories, other researchers had less exalted aspirations and less willingness to plan grand, all-encompassing research programs as they continued to ask questions about differentiation. Moving beyond the earliest cell divisions and developmental stages, they were experimenting with transplantation of tissues and parts from one developing embryo to another. The German embryologists Gustav Born, Hans Spemann, and the German-trained Ross Harrison led the way with this research. Once again, frogs provided a favorite research material because of their availability and large visible eggs. In addition, the variations in color and size among closely related species made them ideal for transplantations. Taking a piece from one frog egg or embryo and sticking it on or into an embryo of a different species allowed the researcher to discover which part dominated. Would the piece develop along the lines that it would have if left alone, or would it adapt to its new surroundings and follow the new pattern? To what extent, in other words, was its fate predetermined?

Gustav Born first introduced the idea of transplantation by taking tissue from one organism, the donor, and transplanting it to another organism, the host. If the two organisms were sufficiently different types, either different species or subspecies, the transplantation was considered heteroplastic; with similar organisms, the procedure was homoplastic or just direct transplantation. Transplantation resulted in a hybrid organism, and frog hybrids proved remarkably resilient in their ability to continue growing, differentiating, and developing even after fairly radical interventions. By the 1890s and into the twentieth century, Ross Harrison adopted the technique and began transplanting bits of frogs' tails from donors to hosts. By splicing on different colored pieces, he hoped to track cell movements during the following developmental stages.[26] At about the same time, Hans Spemann took up similar studies. As they learned more about cell movements and the way that layers seem to develop, Spemann took pieces that normally give rise to eyes or ears, for example, and transplanted them to odd places on the host. Amazingly, they developed into eyes and ears, right there on the back or the belly, or wherever. The transplanted pieces seemed to carry a good deal of specificity with them, yet their fates were not entirely determined by their origin. Under some conditions, the cells adapted to the new conditions as well.

Given what Roux and Driesch had learned about the differences between frog and sea urchin eggs, it would have been valuable if Harrison and Spemann could have carried out similar experiments with sea urchins. What would happen if they transplanted tissue from one urchin to another? Would they see patterns like the donor's, or would the tissue be more adaptive, as the eggs appeared to be, and able to respond to surrounding conditions? Unfortunately, neither Harrison nor Spemann was in a position to work with sea urchins, since they were working in inland laboratories rather than at marine stations. More importantly, however, the experiments would not have worked the same way. Frog eggs are exceptionally tough, set in their ways, and able to continue along their prescribed paths despite interruptions and perturbations. As a result, frogs made a choice subject for heteroplastic transplantation studies of development.

As the studies accumulated, a much more detailed picture began to emerge of how cells and tissues move in the course of differentiation. One beauty of these studies was that they did not have to kill the patient, so to speak. The organism lives and continues to develop, with its transplanted parts and its resulting hybrid form, and the biologist can watch. It is not necessary to kill it, chop it up, slice it, and look inside. Yes, investigators continued to do those things too, of course, and they produced marvelously detailed cross sections of many, many developing organisms. But just watching and recording what happens at each step along the way also yielded valuable information.

The question for us now is: information about what? For scientists of the early twentieth century, the object was to learn about the processes of individual development. Yet we see that they were also very much challenging traditional notions of what an individual life is and when it begins. If it is possible to transplant parts, even at the early stages of embryonic development, and produce a new organism much different from the one that would have developed, what does that mean? If one organism is made up from parts of two, is that life or a double life, or what? A hundred years ago, biologists did not ask these questions just that way. Their research apparently remained sufficiently removed from public awareness and philosophers' reflections that metaphysical questions did not gain wide currency and did not challenge fundamental assumptions about life. Instead, preformationists remained persuaded of their view and felt reinforced by genetics and hereditarian thinking. And epigeneticists remained equally persuaded of their view while exploring and probing to understand life's processes. Once people began to sort through the flood of new data, however, experimental embryology forged a path to deep, new questions.

With the "gold rush" in experimental embryology, as Ross Harrison called it, old questions returned in new ways. What guides development? When is the process fixed, or determined, and when can it regulate in response to changing conditions? Does the organism have some sort of internal organization that determines developmental patterns, and if so what and how should we study it? Several researchers developed concepts of gradients (or graded levels, usually of chemical concentrations) to explain differentiation. Others postulated the existence of fields, while still others pointed to crystals or fabrics as models for understanding development, as historian of science Donna Haraway explained very clearly in her excellent study of these "metaphors of organicism."[27] If the embryo, beginning even in the egg, has gradients of material or forces or densities or something relevant to development, then perhaps those differences cause the differentiation of material and hence produce form from the unformed.

Perhaps the egg already has differences, and there may be different "potencies" or "potentials" laid out in the structure and material of the egg. In this case, Spemann asked, what induces development and differentiation to take place? This question led him to the concept of induction, or the initial stimulation of developmental changes. Most likely, Spemann concluded, embryonic induction is chemically produced—somehow. Yet some material or some embryonic tissue might more easily be induced than other material.

Spemann and his lab group began with pictures of normal developmental stages in mind, since they incorporated the assumptions that the stages are regular and predictable. This approach suggested that each stage, in an important sense, brought something that caused the next stage. In particular, Hilde Mangold (Hilde Proescholdt before she married fellow student Otto Mangold) focused on the upper or dorsal lip of the blastopore and its role in induction for her dissertation under Spemann. This is the stage when the blastopore has opened and cells begin to stream inside the opening. As they move past the dorsal lip, differentiation begins.

Mangold demonstrated the apparent influence of this region by cutting out a piece of the dorsal blastopore lip in a salamander and transplanting it to another embryo, then watching as it "induced" another whole organism in that spot. Demonstrating the special capacity of just that piece of tissue, this experiment also provided suggestions for future research to illuminate the processes of differentiation that followed induction. Yet, according to a fellow student, neuroembryologist-turned-historian Viktor Hamburger, Mangold did not entirely understand the implications of what she had found. Spemann did. As Hamburger concludes, "Hilde Proescholdt, who in the meantime had become Mrs. Mangold, was not happy that Spemann had added his name to her thesis publication, while [Johannes] Holtfreter and I and all the rest of us saw ourselves proudly in print as sole authors. Moreover, Spemann had insisted on having his name precede hers! But Spemann was perfectly right in claiming precedence, while she apparently did not fully realize the significance of her results. It was not granted to her to live to see the great impact her experiment had on the course of experimental embryology." This impact included a Nobel Prize for Spemann for this work, including his interpretations about induction and its implications for differentiation and development.[28]

Spemann concluded from this experiment that the dorsal lip of the blastopore is capable of inducing differentiation of the entire organism.[29] Organization of an organism begins with what he called "the organizer." That may not sound like it explains much, and by itself it does not. But the assumption that there is a material something that initiates the process of differentiation and development of an individual living organism was important.

In fact, Spemann's particular interpretation ran into trouble fairly quickly when Johannes Holtfreter and others demonstrated that quite a number of other materials besides the dorsal lip of the blastopore could also induce development. In fact, even quite dead material could induce the organism. How could this be: what did this say about life and the most fundamental living processes of development? The "organizer" was not quite so neat, nor quite so causally effective as Spemann had hoped. This discovery raised questions about the specifics of his theory and suggested that other analytical lines of research might be equally valuable. Yet the experimental embryological study of organizers, fields, gradients, and internal structure and processes continued fairly aggressively through the 1930s. Plenty of questions remained about the process of induction, whatever starts it: is it a chemical process or mechanical; could there be something like a special "charged" field, or is a matrix—of something—set up in the embryo that later plays out in development? Questions abounded, answers were scarce, but more and more lines of exciting and productive research opened up.

Explantation

A new research approach grew out of transplantation. If it was possible to transplant parts from one developing organism to another, then why not transplant tissue from an organism to a separate culture medium? Ross Harrison pursued this approach by culturing tissues outside the body for the first time. His first attempt involved removing frog neuroblast cells, those that normally give rise to nerves. He placed these in a medium of frog lymph, thinking that this would be a reasonably congenial environment. Harrison asked "what will happen if I do this?" or "if the nerve fiber develops outside the body, then what does this tell us about normal development?" In fact, once he had managed to produce reasonably antiseptic culture conditions, the fiber did develop. It grew out nicely into the surrounding medium and looked, as Harrison interpreted it, just as it would under normal circumstances.[30]

This was marvelous: explanting cells and tissues to grow outside the body gave the researcher access to what normally happens out of view. The cells were reaching out their fibers and growing, right there in broad daylight where it was easy to watch what they do. If this experimentally created process truly paralleled normal development, it was a powerful research tool for embryologists. Indeed it was, though it has taken investigators many decades to realize the potential, to produce techniques and to refine questions sufficiently to take advantage of the possibilities. Tissue culture was even more immediately important for medicine. Though Harrison himself decided to pursue experimental embryological research rather than continue with tissue culture, others carried on the work—for example, Alexis Carrel, who eventually received a Nobel Prize for his contributions. Harrison was considered for a Nobel, but then World War One intervened and the prize was not awarded for political reasons.

Experimental embryology reached a lively intensity of activity in the early decades of the twentieth century. As institutional biology expanded, especially in the United States, more and more biologists joined in the study of development. As medical applications attracted funding, through the Rockefeller Foundation and the Carnegie Institution, as well as at new universities like the University of Chicago, the types of questions asked and research programs pursued also expanded. Embryology had a place as a field in medical schools, where it was useful for the study of causes of pathological problems, miscarriages, and deformities, for example. Experimental embryology also had a home in the biological sciences, as the experimental study of the question first asked by Aristotle: how does the unformed become the formed?

Once investigators outside of Roux's small group in Germany had set aside his and Weismann's particular emphasis on the germ theory, American and British experimental embryologists largely ignored details of inheritance. Yes, the egg is inherited, as is the sperm. That is taken for granted. But little more could be done to study directly the way that heredity leads to development, and assigning much causal value to the genes implied a preformationist view that embryologists rejected. So, as genetics became a field after 1910, experimental embryologists largely ignored it for decades. As late as the 1950s, Harrison at Yale saw little need for a geneticist on the biology faculty. Nor did Spemann see the need in Germany. Yet by the 1960s, their approach, however brilliant it had been, seemed shortsighted. Certainly today, no embryologist would even consider ignoring genetics. What happened during the intervening decades?

Genetics, Embryology, and Cloning Frogs

While experimental embryologists busily chopped up animals to study regeneration, provoked artificial parthenogenesis, transplanted chunks of bodies, and cultured tissues, other scientists were looking at heredity. The embryologists, as noted, did not deny the importance of heredity. They just set it and what they saw as its preformationism aside. Given: the egg cell is inherited; something important is going on in the nucleus and the chromosomes. But we are going to study development and embryogenesis through the process of differentiation because that is what interests us, embryologists might have said.

Edmund Beecher Wilson did not share this attitude, as he demonstrated in his superb study of *The Cell in Development and Inheritance* in 1896 and again in 1900. He waited for over two decades to produce the next edition, however, and it grew —and evolved—considerably. As he noted, so many important things had changed by 1925 that the third edition was really a new book, with a new perspective on cells and organisms. Experimental embryology was one of the important changes.

The Mendelian-chromosomal theory of inheritance was another, and that in turn wound together several threads of research.

At the end of the nineteenth century, expert cytologists, including Theodor Boveri, had made considerable progress in describing details of what the chromosomes do at all stages of development. Of course, Boveri could not just look inside the nucleus while the cell was dividing. This was an area that required painstaking fixing, preserving, staining, describing, and interpreting. Unfortunately, once the researcher had killed a particular cell to study it, that same cell was no longer available for further study. One might say that an uncertainty principle applied: it was not possible both to study the changing cell and also to know what it would change to or from.

Therefore, even though Boveri, Wilson, and others concluded from their observations that there are chromosomes in the nucleus and that these same chromosomes persist through cell division after cell division, others disagreed. They insisted that the chromosomes are made up of temporary coagulations of chromatin (stainable substance) and that the apparent chromosomal structures are ephemeral. These form and then disaggregate, reform in new ways, and so on. There was no way to be sure with the current evidence, though the best approach seemed to involve studying as many cells as possible, over as many different times as possible, and collecting, synthesizing, and comparing data. If the chromosomes really were fundamental for heredity and in some way for guiding development, there was much at stake in studying them. This research ultimately became the foundation for our current understanding of the genome.

Boveri's view, reinforced by Wilson and his students, was that the chromosomes are autonomous and defined and that they maintain their separateness throughout the life of the organism.

They divide during cell division but nonetheless remain autonomous and distinct. Boveri argued for this principle of the "Continuity of the Chromosomes" and the principle of the "Individuality of Chromosomes" in his work on the nematode worm found in horses, *Ascaris megalocephala*. This was an excellent medium because it had a small number of relatively visible chromosomes.[1] Still, not everyone was convinced, some because they did not agree that the chromosomes had much importance. Boveri needed more support for his interpretation.

Mendelian heredity provided just that. Gregor Mendel, the remarkable monk working away in the monastery gardens cultivating the sweet pea, has received considerable attention in the history of science. Some have elevated him to near-saintly status, while others have sought to detract from his importance by pointing out that he did not actually persuade anyone of his theory during his own lifetime. A few detractors have insisted that Mendel's data look so good statistically that they must have been fudged.[2]

Only in 1900, after having been ignored for thirty-five years, was Mendel's work "rediscovered" by three different researchers. Even though Mendel's initial paper had never really been "discovered" in the first place, it now fell on fertile soil. The context had changed since its initial publication in 1865, and its ideas immediately riveted the attention of biologists. Even those who did not accept Mendel's argument completely saw possibilities in it. Mendel had concluded from his studies that at least two regularities occur in inheritance. He made the background assumption that parents pass on through heredity something material and defined, rather than vitalistic or amorphous. Mendel did not pretend to know what that something was in his few published papers. Rather, he said, each offspring begins with "something" from the father and "something" from the mother.

WHOSE VIEW OF LIFE?

The "somethings" are "factors" that correlate with traits, Mendel concluded, occasionally with one factor per trait but generally with more factors involved.

Mendel's remarkable discovery was that, given a collection of organisms, one could count the distribution of variations of a trait and discover regularities in the proportions of its expression. Take the results of hybrid peas made from crosses between two varieties that have clearly visible differences in a defined number of characteristics. Mendel found seven characteristics, or traits, that revealed predictable regularities and observed the proportion of individuals with each of the variations for each characteristic. If he were looking at the pea coat, for example, he noted the proportion of those with wrinkled and those with smooth coats. Compiling the data, he could predict how many would have yellow or green peas, particular color flowers, and so on. These quantitative summaries could predict when a cross would produce one-quarter of the offspring with a particular trait, or one-half, or three-quarters. That is, his studies gave him the statistical and empirical capacity to predict the proportion of each variation among the offspring in each of many successive generations.[3] It was possible, for example, to cross red and white varieties in some organisms and get not pink offspring but all red. Or to cross a variety in which all the individuals carry only what Mendel called a "dominant" factor for a characteristic with a variety carrying what Mendel called a "recessive" factor for that characteristic and to get offspring all with the dominant characteristic—like color or size or shape. But crossing hybrids in which each individual has some mix of dominant and recessive factors will lead to something mathematically quite interesting.

Mendel's seven characteristics did not include flower color but rather details about the seeds and pods, or stem length or flower position. His first example compared the shape of ripe seeds: were they smooth or wrinkled? Cross hybrids of these two varieties, and the next generation would yield a ratio of approximately (very approximately with real plants, since nature is, in fact, rather messy) three offspring with the dominant characteristic to every one with the recessive. Cross those offspring among themselves, and Mendel thought that the next generation must yield a ratio of 1:2:1 among different inherited combinations of factors.

But what does this mean? How can the ratio be 1:2:1 when there were only two factors corresponding to two different varieties of that particular characteristic to start with? This is Mendel's insight: that two factors can, in fact, combine in four different ways, two of which yield the same result. The possibilities are a cross of (1) a dominant factor from the mother with recessive from the father, (2) dominant from the mother with dominant from the father, (3) recessive from the mother with dominant from the father, and (4) recessive from the mother and recessive from the father. If the characteristic is not sex-linked, and since dominant dominates, every time a dominant is contributed the result will be a dominant characteristic. Therefore, the four possible outcomes are (1) dominant, (2) dominant, (3) dominant, and (4) recessive. With equal probabilities for each match, the offspring will include three plants with the dominant for every one plant with the recessive characteristic. But notice, as Mendel did, that the dominants are not all the same: (1) and (3) are really hybrids themselves. They appear as plants with dominant characteristics, but hidden away inside are factors for both dominance and recessiveness. In contrast, (2) is a "pure" dominant, just as (4) is a pure recessive. The difference has implications for what will happen in the next generation,

Mendel recognized, and it was his ability to make predictions that makes the Mendelian view so powerful.[4]

His studies led him to what were later termed Mendel's two laws. The "law of segregation" held that the offspring inherits a "factor" from each of its parents and that these factors segregate. That is, they will not blend or merge, but rather remain separate. Perfectly segregating factors for red and white characters would not make pink, therefore, but only either red or white. This is a conservative view that explains why offspring are like their parents or grandparents and not something new, completely different, or in between.

Second, what was labeled Mendel's "law of independent assortment" held that for each separate characteristic, the factors remain separate and not linked to each other in any way. That is, factors for pea coat wrinkledness or smoothness sort completely separately from those for flower color or other characteristics. The factors were more like separate beans in a bag than like the linked beads on a rosary—not that Mendel ever put it just that way, but his interpreters have.

What so excited biologists after 1900 about Mendel's ideas were the possibilities for explaining both heredity and development through this predeterminist view. It had already made clear that what is inherited is an egg, the egg is a cell, and this egg cell is highly structured. Furthermore, the egg cell is fertilized by a sperm cell, which brings inherited material from the father in the nucleus and the chromosomes (since the sperm cell is very little more than nucleus and chromosomes). What if the inheritance actually involved transmission of material particles, or factors, of some sort from the parents to the offspring? It would be very convenient if those factors retained their independence and solidity during the cell divisions and into the future embryo and the organism that results. These inherited factors could serve as causal connections across generations and might even explain development through the conservativeness of heredity. Different combinations of factors could explain the range of variations in a family or a population, though not the origin of new variations. The idea was attractive, and evidence that it was true, at least for one species of sweet peas, at least for some characteristics, held promise as an explanation of heredity. Exploring the possibilities seemed a very productive research program indeed.

Boveri clearly saw the prospects. As biologist and historian Henry Harris puts it, Boveri "could not avoid noticing that the stability and individuality of the inherited traits described by Mendel could be readily accommodated by the chromosome mechanics that he had done so much to clarify. He did not again doubt that it was the behavior of the chromosomes that determined the rules of inheritance Mendel had discovered."[5] From 1902, Boveri worked on demonstrating that connection and articulating a Mendelian-chromosomal interpretation of heredity that would bring together both lines of research and thinking.

In retrospect, the early 1900s seems to have been a critical time. Surely, it might seem, any rational biologist should accept the beauty of the theory and see how neatly it fit the evidence. Yet it was not perfect. Recall that the Mendelian interpretation called for independent assortment, for example. How could factors sort independently if they were lined up on chromosomes and if chromosomes retain their individuality and do not fall apart and recongeal with every cell division? Why did only a few select characteristics follow the prescribed pattern? The theory provided a framework, but much work remained to be done. Furthermore, it was still reasonable to reject the interpretation. Thomas Hunt Morgan resisted, for example, and this is especially interesting because he soon became one of the strongest supporters of the very theories he first rejected.

Through early 1910, Morgan continued to assert what he thought was wrong with the Mendelian-chromosome theory in ways that summarized his thinking of the previous decade. Morgan rejected what he saw as the unjustified preformationist reasoning inherent in the theory. In addition, he was driven by concerns that the theory relied heavily on hypothetical and unobserved hereditary particles or factors. Morgan very much saw the egg as a balance of chemical and structured material that is influenced by heredity but primarily plays out in response to environmental conditions. His study of sex determination had shown him that.

How does an individual organism become one sex or the other; what makes one become a male or female and when does it happen? Traditional interpretations had pointed to external factors: perhaps the water supply, what the mother ate, or other environmental conditions the mother experienced at conception or at a critical stage of gestation determined the sex. By 1903, Morgan had ruled these out as not consistent with the data, though others had not. Yet he remained convinced that it was the balance of internal factors that determines the outcome. The egg and early embryo hold multiple possibilities, and "here, as elsewhere in organic nature, different stimuli may determine in different species which of the possibilities that exist shall become realized."[6]

Even when evidence accumulated that sex might be linked to the presence or absence of particular chromosomes, Morgan pointed out that there was a lot of room for additional interpretation about just how that works.

Mendelian preformationist factors could provide an explanation of heredity, Morgan agreed, but with considerable limits. He concluded in 1909 that "the preformation idea has always led to immediate, if temporary successes; while the epigenetic conception, although laborious, and uncertain, has, I believe, one great advantage, it keeps open the door to further examination and re-examination. Scientific advance has most often taken place in this way."[7] Morgan saw epigenesis, under the current circumstances, as better science.

Furthermore, Morgan rejected reliance on hypothetical entities. As he wrote in the same article, "In the modern interpretation of Mendelism, facts are being transformed into factors at a rapid rate." He lamented, in other words, that "we work backwards from the facts to the factors, and then, presto! explain the facts by the very factors that we invented to account for them."[8]

In early 1910, Morgan wrote another, longer paper arguing the same points. While it was acceptable to pursue the Mendelian and chromosome theories that relied on inheritance of material particles that cause development and to see where they would lead, he much preferred a "physico-chemical reaction" that could respond to changing environmental conditions. Indeed, it was vitally important to remember the highly problematic status of both theories. Thinking in terms of particles might hold allure because it seemed simpler or more "picturesque or artistic," but in science we should not follow a theory for these reasons alone, nor simply because it becomes popular, Morgan urged.

Sometimes a theory like the reaction theory, which is more provocative and offers more "restlessness of spirit," is "more in accord with the modern spirit of scientific discovery." To do good science, which Morgan believed was possible and certainly desirable, "Whichever view we adopt will depend first upon which conception seems better in accord with the body of evidence at hand concerning the process of development."[9] Morgan was quite clear about this. Yet a few months later, he was writing in favor of Mendelism and chromosomes both as the bearers of heredity. Why?

Morgan was adaptable, willing to follow what he saw as the best science. Early in 1910, he discovered something that convinced him that exploring Mendelism was the best science after all. Namely, he found a white-eyed male *Drosophila* fly. It is difficult to underestimate the power that this tiny fruit fly had in shaping the course of modern studies of heredity and even how we think about ourselves.

Morgan was an opportunist in the best sense of the word. He studied whatever organisms he could, using whatever legitimate methods and asking whatever important questions, as long as they advanced research that was consistent with the best available science of the time and yielded new knowledge. As a result, he and some of his students were studying fruit flies as well as other organisms in his laboratory at Columbia University. They were recording what they saw: what variations, what combinations of traits, what proportions of variations in a population—the sorts of things that the Mendelian program called for studying. They continued to find the "wild type" with normal red eyes and other standard patterns. Then, one day in 1910, he saw a white-eyed male.

What was important about this event was not that Morgan saw the one white-eyed male and recorded it but that he recognized, as many would not have, that this particular fly did not fit the usual pattern. He saw the value in the different and the surprising. In particular, he recognized the potential for future study, using the sorts of cross-breeding that Mendel and others had developed in agriculture. Morgan took that one fly and bred it. Then he carefully recorded the results of the cross between the white-eyed male and his red-eyed sisters. His short paper, less than three full pages, recorded that 1,237 offspring had red eyes and only three—all males—had white eyes. These three, which Morgan assumed resulted from further "sporting" or mutation, he ignored. This in itself is fascinating. When he saw one white-eyed male amid a mass of flies in his laboratory at Columbia University in Manhattan, he recognized it as important and potentially valuable and singled it out for further study to ask "what happens if . . ." it was crossed in various ways.

Now that he had done the first cross, the appearance of a tiny number of additional white-eyed males was no longer interesting and could be ignored. Morgan had a genius for recognizing what was most important for science—possibility and the opportunity to ask questions and develop answers and explanations that could be tested and developed in order to address more and more questions. Never settle for too simple an answer if another will take you further in the long run, was his message.

With his flies, Morgan did ask further questions. Since the first hybrid cross produced virtually all red eyes, what would happen in the next cross—among those red-eyed hybrid offspring? Here he got 2,459 red-eyed females, 1,011 red-eyed males, 782 white-eyed males, and not a single white-eyed female. The particular numbers do not matter for us, but they show the detail that Morgan provided. And they show the clear way he thought through the problem. While breeders had been carrying out crosses and creating hybrids of many sorts for centuries, only with Mendel and his rediscovery did it seem important to keep count of exact numbers within large populations. (Counting lots of tiny fruit flies is not an easy nor a particularly romantic task, as much of basic science is not.) To be motivated to do the research, these biologists had to believe that the numbers and the ratios they make up actually matter.

Morgan continued his counting and his crossings. The flies proved an excellent research subject, since they breed fast and

WHOSE VIEW OF LIFE?

therefore provide many generations in a short amount of time. Many flies fit into a small space. They are reasonably cheap, eating rotting bananas, for example. Flies brought an added bonus: the ubiquitous bananas hanging in the window of the fly lab also attracted other biologists—cell biologist Edmund Beecher Wilson and others reportedly wandered in and out of the lab quite regularly. Certainly, working with flies offered many advantages over breeding pigeons, as Darwin or as American embryologist Charles Otis Whitman at the University of Chicago had done. Indeed, one raw winter day in January, Whitman went out to care for his pigeons because a sudden storm threatened them, and he developed pneumonia and died as a result. *Drosophila* flies provided a very attractive alternative research subject.

Drawing on his accumulating observations, Morgan developed hypotheses to explain the results. The eye color appeared to be "sex-limited," he concluded, and therefore perhaps was caused by a "factor" for white eyes that was carried on the male chromosome only, whereas the red eye factor was carried by all females. White-eyed males occur only as a result of a mutation, he reasoned, when a male for some reason does not carry the factor for red eyes that he would normally receive by heredity from his mother.

Morgan reasoned through, step by step, the possible interpretations based on inherited material particulate factors—the very same factors that he had argued so energetically against only months before. Follow the evidence, Morgan always insisted, and pursue productive lines of research. Follow wherever they take you, even to inherited factors and to what became decades of crossing fruit flies by a laboratory team of researchers that historian Robert Kohler called *Lords of the Fly*.[10]

What Morgan concluded in that important little 1910 paper was that the eye color seemed to be linked to the sex, which meant that females could carry the cause of the mutant white and males would exhibit the trait in greater numbers. But there was already evidence that sex was linked to chromosomes. Therefore, if eye color was linked to sex and sex to chromosomes, at least it made sense to pursue the hypothesis that eye color was linked to chromosomes. Figuring out how opened rich research possibilities, and Morgan led the way in pursuing them.

With further study of chromosomes, characteristics, mutations, and populations, the chromosome theory of heredity became well established. Accordingly, chromosomes were seen as carrying the material of heredity from one generation to the next, with factors lined up along the chromosomes. Morgan's students undertook a series of detailed and painstaking studies that allowed them to begin "mapping" the factors, named genes, on the chromosomes. They could not literally see the factors or map them visually, but rather they used statistical analysis. With what frequency did any two characteristics occur together in the organism? If the frequency was high, they concluded that the factors must reside near each other; if low, then the factors must be farther away or on different chromosomes altogether.

This exercise, taken as confirming the Mendelian-chromosomal theory of inheritance and placing it on solid ground, actually undercut Mendel's second "law." In fact, this line of research was predicated on the assumption that there is *not* the independent assortment that Mendel had seen. Indeed, by thinking contra Mendel and predicting that linkages would occur, the investigators could use the frequency of those linkages to map proximity on the chromosome. Yet, this did not mean that Mendel was wrong. Rather, it began to look as if Mendel had chosen to study seven different and very special characteristics, with factors located on several different chromosomes. That required a certain genius of observational insight on Mendel's part, as historians have pointed out. Mendel looked beyond the messy confusion of variations in nature and saw patterns and regularities. Morgan's group could take the next step and begin to accept some of the variation and make sense of it too. By doing so, they placed genetics at the center of twentieth-century biology.[11]

During the next decades, as several historians have documented persuasively, researchers followed various lines of investigation to discover what is inherited and how inheritance works. Was it purely a chemical interaction, such that whatever was inherited acted chemically on the surrounding cell cytoplasm to effect differentiation? Or was there also a physical or mechanical pushing or pulling, like some sort of force or vector? What was the chemical makeup and the structure of the inherited material that made up genes?[12]

A half-century of research culminated in 1953, when James Watson and Francis Crick's one-page paper announced the structure of DNA. Promising more detail elsewhere, the authors casually mentioned at the end that "it has not escaped our notice that the specific pairing we have postulated immediately suggests a possible copying mechanism for the genetic material."[13] Those possibilities, and the implications for the importance of heredity in defining life, shaped much of the search to identify the material of heredity and much of the rush to spell out the implications of DNA structure. The presentation, as well as the discovery, of genes as an almost magical stuff of heredity contributed to a recurrence of biologists' emphasis on inheritance and its great influence on the development of life.

Hereditarianism and Eugenics

Of course people thought about heredity before Mendel. Henry VIII obviously cared a lot about heredity! So did the royal families affected by hemophilia, Charles Darwin, parents concerned about their children, and so on. George Bernard Shaw, in his 1903 play *Man and Superman,* showed how the discussion of heredity changed after the rise of genetics around 1900. Shaw's characters want to produce supermen by breeding effectively. Do not waste superior genetics on inferior partners, is part of the message.

A modernized Don Juan story inspired by Nietzsche and the general British intellectual climate, this comedy suggests that we should reject prudery and conventions and just breed to make good bodies and watch the race evolve. As Shaw's "revolutionary" Jack Tanner wrote in the "Revolutionist's Handbook" at the end of the play, "The changes from the crab apple to the pippin, from the wolf to the fox to the house dog, from the charger of Henry V to the brewer's draught horse to the race horse, are real; for here Man has played the god, subduing Nature to his intention, and ennobling or debasing life for a set purpose. And what can be done with a wolf can be done with a man. If such monsters as the tramp and the gentleman can appear as mere byproducts of Man's individual greed and folly, what might we not hope for as a main product of his universal aspiration?"[14]

This very Shavian offering is probably not typical of what most people were thinking at the turn of the twentieth century. At the same time, however, new ways of thinking about heredity and breeding were gaining wider attention. Eugenics was the major public expression of the mood. Inspired by Francis Galton in England, this movement played out differently in Germany and in the United States. The eugenics movement shaped much of our current thinking about heredity and development and helped shape our social intuitions about morality in relation to human reproduction and heredity. It is therefore worth recalling the underpinnings that made the social eugenics movement of the early twentieth century possible.

The temptation to try to control destiny is perhaps overwhelming. Why leave one's future, including one's offspring and inheritance, to fate if selective breeding could ensure a better future? That was, in effect, Shaw's message: consider individual choices and the impacts of those choices in individual breeding. Look beyond the individual as well. What about the destiny of the society? Can we make the German race more "pure" and improve it? Hitler answered that he could, or that he could use that argument to try to justify his obviously unjustifiable extreme actions. Can we improve public health in the United States by removing some of the "defective" members of the population—or rather by removing them from future generations by preventing them from breeding, American eugenicists asked, and they too concluded that they could.

Eugenics seemed to make perfectly good sense: good policy based on good science.[15] If some people are better genetically than others, let us encourage the best to have more children and improve the population through "positive eugenics." Let us also sterilize the weakest members and prevent them from adding to the population; in other words, we can promote "negative eugenics" through selective sterilization. Historian Daniel Kevles has documented the decisions of this sort that Americans made and their impacts, state by state, and Philip Reilly explores the legal side of the decisions.[16] Sinclair Lewis's *Arrowsmith* provides a contemporary poke at the eugenics craze, with a passing story about the "eugenical" family being celebrated at a local fair and known to be a bunch of misfits and thugs, not even a family at all. The search for the perfect family, perfect health, a perfect future with perfect babies, and perfect solutions to problems—all through science and social implementation of public health laws—was alluring. It was, obviously, also based on shaky assumptions and fraught with problems.

States passed eugenics laws, and some of the states used the laws to sterilize large numbers of people. By one count, nearly 61,000 Americans had been sterilized by 1958, over 20,000 in California alone.[17] One example tells us a great deal about the thinking of the time. The case of Carrie Buck reached the U.S. Supreme Court, whose 1927 ruling upheld the authority of the state both to sterilize and to interpret what counts as evidence in favor of the claim that somebody is an imbecile.

The state of Virginia passed a statute allowing for sterilization of "mental defectives." Most often it was residents of the State Colony for Epileptics and Feeble Minded who were targeted, and hundreds were ordered sterilized. Carrie Buck was one of these. A young mother herself, she was the daughter of Emma Buck, a poor and uneducated woman who had been institutionalized and later scored poorly on the relatively new Stanford-Binet IQ test. Emma's unmarried daughter Carrie, who was taken to the state hospital when she became pregnant, was judged by the IQ test also to be an "imbecile"—a technical term associated with particular test scores. Carrie Buck had been placed in a foster home when her mother was institutionalized, and it became clear only much later that she was raped while at that home. The foster family took young Carrie to the State Colony when she became pregnant. With an illegitimate child and uneducated like her mother, she was officially declared to be a candidate for sterilization under the new eugenics laws. Much hinged on the judgment of the mental status of her daughter, Vivian.

In one of his most compelling and poignant essays, and with help from other researchers, Stephen Jay Gould showed that the entire evidence in favor of the claim that this third generation was mentally defective—the all-important third generation that could establish a pattern—came from a single social worker. The evidence from that social worker, Miss Wilhelm, who had examined the baby Vivian Buck at seven months old in the mental hospital, was only this: "It is difficult to judge probabilities of a child as young as that, but it seems to me not quite a normal baby. In its appearance—I should say that perhaps my knowledge of the mother may prejudice me in that regard, but I saw the child at the same time as Mrs. Dobbs' daughter's baby, which is only three days older than this one, and there is a very decided difference in the development of the babies. That was about two weeks ago. There is a look about it that is not quite normal, but just what it is, I can't tell."[18]

From our perspective, Miss Wilhelm's assessment of Vivian's "imbecility" was manifestly weak and highly problematic. At the time, however, it was not seen that way, largely because of the dominance of hereditarian thinking and the public health emphasis that had been placed on heredity. Supreme Court Justice Oliver Wendell Holmes declared confidently, in a landmark ruling, that Carrie Buck "is the probable potential parent of socially inadequate offspring, likewise afflicted, that she may be sexually sterilized without detriment to her general health and that her welfare and that of society will be promoted by her sterilization." Therefore the Court upheld the decision to sterilize and judged that the law had been properly applied. Furthermore, "We have seen more than once that the public welfare may call upon the best citizens for their lives. It would be strange if it could not call upon those who already sap the strength of the State for these lesser sacrifices, often not felt to be such by those concerned, in order to prevent our being swamped with incompetence. It is better for all the world if, instead of waiting to execute degenerate offspring for crime or to let them starve for their imbecility, society can prevent those who are manifestly unfit from continuing their kind. The principle that sustains compulsory vaccination is broad enough to cover cutting the Fallopian tubes. *Jacobson v. Massachusetts,* 197 U.S. 11. Three generations of imbeciles are enough."[19]

For Holmes, sterilization was like vaccination, something that really is not so bad for anyone and that is done for the sake of protecting the larger public health. There is a perspective from which Holmes's judgment seems a reasonable and justified restriction on individual rights for the sake of protecting society. But it only seems that way if, first, we are quite sure we understand the danger being addressed and, second, we clearly will be protecting society from it through our actions. By sterilizing people like Carrie Buck we must, in fact, be protecting future generations from a certain danger. And we must be quite sure of our evidence in favor of the claim that there is a danger. The claim that Carrie Buck's ability to have children would be a danger to society was grounded on the evidence that she had, like her mother, had an illegitimate child, that as an uneducated young woman she could not do well on a test, and that a social worker thought her young baby looked odd. Weak grounds, we see now.

Naturalist Stephen Jay Gould insists that we should have seen the weakness even then. Even more importantly, we should not have allowed the Court's reasoning to reach beyond the case at hand. The even worse tragedy of this particular case, Gould documented, was that Carrie's sister Doris was also sterilized. Doris had never had a child, illegitimate or otherwise. She was not tested. Rather, hereditarian thinking led to the assumption that since she was also one of the three generations, she was also de-

WHOSE VIEW OF LIFE?

fective and a burden to society. It was deemed just and reasonable to sterilize her, too. She was never told the nature of her operation, and only much later as historians studied the case did they realize why Doris never could have the children she so sorely wanted. The state of Virginia, and eugenic policies generally, had taken away her future offspring.

Is the Bucks' case a tragedy? Certainly, yes. Is it a gross miscarriage of justice? Certainly, yes, in at least one sense. Yet the law was upheld, so the case stands even more securely as a miscarriage of justice and a misinterpretation of what should count as sufficient evidence. The decision was plausible, and indeed applauded, because of the strength of hereditarian thinking, along with the assumptions that it was possible to know what was defective and that we could do something about it. That hope to control destiny—or to control heredity, and life—led us to bad science in the form of distorted eugenic assumptions, to bad policy, and to bad implementation. By now, with the experience of the Human Genome Project, we ought to have learned not to make similar mistakes. Shame on us if we have not.

It is instructive to note that the United States moved away from the legal force of eugenics more for political than for scientific reasons. Yes, there was increasing awareness of the difficulty of establishing beyond any scientific doubt what counted as "feeble-minded" or "defective." This was a scientific argument. There was also a growing awareness, with World War Two, that the Nazis were trying to justify political actions with eugenic reasoning about "race hygiene" that Americans should not condone. This was a political argument. In addition, there were social and moral shifts.

Whereas Americans in the early twentieth century embraced regulation and public health measures, after World War Two the country began to have greater concern for individual liberties and civil rights. We argued during both the first and second world wars that protecting individual rights and "making the world safe for democracy" was the reason we were fighting, and increasing prosperity in the postwar years shifted attention to what individuals could gain. Public health remained a concern, of course. Vaccination programs and premarital screening for venereal diseases continued to seem like sound policy. But overall, the country was moving toward a classical liberalism focused on rights and individuals. Emphasis was put on public health efforts to protect individuals while public policies such as quarantines and ordinances prohibiting spitting on the street lost favor.

Rather than design eugenic programs to improve populations, we began asking how we could improve the health and heredity of an individual baby—our own babies. Dr. Spock's child-raising guides for parents, combined with campaigns promoting balancing food groups, prenatal regimens for mothers, and the medicalization of more and more procedures, including childbirth, seemed to promote healthy babies and happy families. Prior to World War Two, more mothers than not, even in the United States, still had their babies at home rather than in hospitals. This was partly for reasons of cost, and partly because it seemed a perfectly natural thing to do. But prosperity and progress after the war quickly moved childbirth into hospitals. Sanitation, sterilization, and the most advanced of technologies would ensure the health of the new bundle of joy—for the growing working, middle, and upper classes, at least. Life was good. We could produce good lives for our families and ourselves. The "medicalization" and "professionalization" that apparently made progress possible also eventually brought about an important shift toward a reliance on expertise, as well as toward arguments for individual right of access to expertise, even at public expense.

In the early twentieth century, the physician, poet, and story-writer William Carlos Williams had reflected on the trend to count on experts to solve our problems, especially in medicine. He worried over it, but he did not try to stop it. Writing about a poor little "brat" in the hospital during the Depression, he pondered what medicine can and should do for a small child. He could patch her up and send her back to poverty and malnutrition, to a mother who did not want and did not have the capacity to support a child. In the end, little Jean Beicke did not get well, even though the doctors tried. An autopsy revealed the pathological reasons for their failure, and it is clear that Williams really wanted to heal her after all. At least he wanted to want to heal her, but he knew he could not address the underlying social problems as he could the medical ones.

In "A Night in June," Williams considered how much medicine is right. A patient of his was delivering her ninth child, at home, and Williams fell asleep while waiting. He began arguing with himself. How much science is right, and how much should a physician let nature take its course—in this case, let the mother do her job.

We can know a lot of science and medicine, Williams makes clear, but we have to work a lot harder to discover how and when to use it. He lamented that we had—well before the 1960s when he published the stories that he had written through the previous decades—already come to rely too heavily on the science, too often leaving the humanity behind. "Science, I dreamed, has crowded the stage more than is necessary." A good life cannot come from science alone. Yet without science, things go wrong, as the doctor well knew. The woman's first child was "a difficult forceps delivery and I lost the child, to my disgust; though without nurse, anesthetist, or even enough hot water in the place, I shouldn't have been over-blamed."[20]

Undoubtedly, tensions exist between the so-called purity and objectivity of science and the political and social role of expertise in delivering "goods" to real people. That tension came with pressures to generate and apply new knowledge to solve more medical problems. Whether, under what conditions, and who should have rights to access the resulting benefits still remain matters of considerable debate. Different pressures apply when scientific research is publicly or privately funded or, as is more commonly the case, some combination of both.

Genetics and Embryology

Genetics attracted considerable public interest and interpretation through eugenics. In addition, even as the scientific study of genetics drew more and more researchers and funding in the 1910s and 1920s, the gold rush continued in experimental embryology. Yet most researchers chose one path or the other. In contrast, T. H. Morgan pursued both. Once again, his work provides valuable insight into the relations between studies of development and heredity.

As the leading lab director for *Drosophila* genetics at Columbia University and later Caltech, Morgan might be expected to have focused on genetics for the rest of his career. Indeed, he and his students wrote the leading textbooks in the field through the 1910s and into the 1920s. For Morgan, however, heredity was only part of the story of biology. There is also development. Perhaps most enticing was the question, what is the relationship between heredity and development? Morgan's 1927 textbook, *Experimental Embryology*, is worth quoting at some length to understand exactly the language this leading founder of genetics was using.

Morgan wrote:

> The arrangement of the genes in the chromosomes bears no relation at all to the arrangement of the parts which the structures of the fully formed individual bear to each other. The developmental changes that we see and try to explain are supposed not to be primarily due to changes in the chromosomes (for the genes remain, we think, intact and perhaps unaltered throughout embryonic development), but in the cytoplasm of the egg. These changes in the cytoplasm are relatively gross processes in comparison with the minuteness of the genes that are the ultimate agents behind them. In the study of embryonic development we see only the gross events; the presence of other agents is inferred from a different kind of evidence. How, then, is the ordered sequence of events, that takes place in the cytoplasm, related to the activity of genes in the chromosomes if there is no correspondence of arrangement in the two? Is it due, for example, to the sequence in which the genes become active? We dare not make any such assumption, tempting as it might be to do so, for there is no evidence that there is any such sequence of activities in the gene.[21]

Morgan went on to note that either the developmental activity in the cytoplasm is directed by the sequence of genes or it is not, but there is no way to tell which. There is no evidence that definitely supports one interpretation over the other. Therefore, biologists need to work from "both ends," by studying the way that genes work and their influences on cytoplasm.

His 1927 book focused on embryology, just as others had concentrated on genetics. In 1934, Morgan brought the two to-gether—at least in the title—in *Embryology and Genetics*. The second chapter, only seven pages long, addressed "Development and Genetics." There Morgan summarized knowledge about in-heritance and concluded with what today seems surprising from a man who received a Nobel Prize for his work on genetics. He said, "The central idea of genic balance is that all the genes are acting, and what is produced is the sum total of their influence. If only one gene is changed, i.e., mutates, the product is changed in some degree, certain organs being more affected than others; but still all the genes are concerned. In other words, the new gene acts only as a differential. This formulation gives a consis-tent picture of the end-products of the genes, but it is quite inad-equate to explain the sequence of changes through which the embryo passes."[22]

Today we have come to see genes as little packets of material inheritance that, in effect, very much direct and even determine what characteristics the organism will have. We know that most characteristics are the product of multiple genes, but there is still a widespread popular sense that for at least some traits, genes are destiny. This is just the sort of interpretation that feeds eu-genic thinking. There is so much temptation to think this way because these genes-as-little-packets-of-heredity seem to pro-vide, as Morgan well recognized, an explanation of what occurs in development. It is very much more difficult to accept that there is much more to learn about how development works and what complex set of influences shapes the organism. Yet we ought not accept a simplistic explanation just because it is avail-able if it is not justified by the science. The jury is still out on how much of development is directly "caused by" genetics.

Despite the allure of genetics, Morgan still saw development in epigenetic terms. Even if there were genes, inherited and ready

WHOSE VIEW OF LIFE?

to jump into action at the right time, it was the total complex of genes that would decide what any particular individual organism would turn out to be like in any particular environmental set of conditions. Research from experimental embryology supported that interpretation, and so, Morgan thought, did genetics. Since all the cells contain all the genes, and since all the genes on all the chromosomes continue to grow and divide so that each new cell gets its full complement, it must be something other than the genes themselves that cause some to become active at any given time and others to become active at another time.

Morgan was right, of course. It would have been easier to have adopted a more strictly Mendelian interpretation, concluding that one gene determines one trait—or rather one set of genes from each parent does so. It would also have been tempting to infer that the genes become active along the sequence in which they are lined up on the chromosome. But assuming it to be true without evidence would not be good science and would not be justified. Morgan urged a more complicated, and what he saw as a more realistic, view of life. He was not alone, but those who took this view needed something to make their proposed research programs productive.

As long as genetics and chromosome studies on the one hand and experimental embryology on the other continued to look so exciting and to attract so many researchers, embryology and genetics would remain largely separate studies. Others might want to do something like developmental genetics, but they did not see how to go about it. Perhaps they had hoped that Morgan would lead the way. He reported that he was roundly criticized for not doing so in *Embryology and Genetics,* but he responded, in effect, "did I not do just what I promised?" The book includes embryology and it includes genetics. It did not, however, introduce anything like embryological genetics or genetic embryology. That was left for others, at a later time, when different lines of evidence and reasoning could be drawn on.

The efforts to study genetics on the one hand and development on the other were concentrated in Germany and the United States. Germany had the tradition of cytological study, and German researchers continued to lead the way with the most advanced staining, microscopic techniques, and advanced experimental methods. Well into the 1920s, many of the Americans who became leaders were trained in Germany or with Germans at the Naples Zoological Station, or were trained by those who had studied there themselves.

In part, the organizational structure of German and American biology contributed to their leadership at various times. Even after the unification of Germany in the mid-nineteenth century, the German academic community was dispersed at research institutes across the country. This meant that Breslau, Berlin, Munich, or Würzburg might each have a leading cytologist or embryologist at any given time. These men might work on subjects, ask questions, use methods, and seek priority in overlapping fields. They were competitive and yet able to coexist in closely related areas. Rather than having some central authority decide that one researcher would study, say, embryology through transplantation in frogs while another explored cell lineage in ascidians and another nuclear inheritance in sea urchins—or whatever—they could decide largely for themselves. This led, I believe, to a healthy interaction among researchers who remained informed about each other's work. This vibrant research community largely broke down with the pressures and distractions of World War One.

In the United States, the 1890s and early 1900s saw the rise of new research universities, like the University of Chicago and

Stanford, and the transformation of others, like Yale and Columbia. Funding from private foundations, led by the Rockefellers and Carnegies, and public funding through the National Research Council and the National Institutes of Health, starting in the 1930s, stimulated the development of biological and biomedical sciences.[23] Graduate programs produced Ph.D.s, who then took their places carrying out more research. The expanded pool of researchers, trained from a few centers, grew into an interconnected group of biologists who communicated through new journals and who both collaborated and competed in productive ways.

Cloning

So, how would it be possible to do a robust study of development *and* genetics? Spemann had a suggestion. Why not perform "an experiment which appears, at first sight, to be somewhat fantastical"? Jacques Loeb had already shown in sea urchins and Spemann in newts that it was possible to start cell division and development in a piece of egg that has no nucleus through a sort of artificial "fertilization" using another nucleus from a later developmental stage. Spemann had done this by constricting the egg, so that one part did not have the nucleus. It remained inert. But at a later stage he allowed a nucleus to pass through a sort of "protoplasmic bridge" and thereby stimulate what would have been the normal processes of division. His experiment raised the possibility of a new approach to transplantation. Why not transplant a nucleus from various developmental stages to an enucleated egg? Then it would be possible to watch each cell division and see what happens next, as was done in the transplantation experiments that had become so familiar by then.

But how to accomplish this? As he reported in his Silliman

Lectures delivered at Yale University in 1936, Spemann imagined that obtaining a "clean" nucleus would not be difficult, that he could simply "grind" some cells between two microscope slides. That would remove the jelly coating that normally surrounds the egg in the earliest stages. A similar procedure would also remove the cytoplasm surrounding the nucleus and leave the bare nucleus, Spemann predicted. That would produce the donor material: the nucleus.

As to producing the hosts, or the eggs without their own nuclei, Spemann did not yet see how to do that. If he had, he probably would have tried the experiment himself. Instead he wrote that for "the introduction of an isolated nucleus into the protoplasm of an egg devoid of a nucleus, I see no way for the moment." Yet, he continued, "If it were found, the experiment would have to be extended so that older nuclei of various cells could be used. This experiment might possibly show that even nuclei of differentiated cells can initiate normal development in the egg protoplasm. Therefore, though it seems an anticipation of exact knowledge to say that 'every single cell possesses the whole apparatus of potencies,' (Peterson, 1922, p. 116), yet this opinion may be right."[24]

So many writers have interpreted this suggestion of Spemann's as foreseeing cloning and even stem cell research that it is worth looking further at the context and significance of Spemann's ideas. While we can point to this passage on one page of a long and important book, there are many others that do not in retrospect seem so full of valuable foresight. What, then, was Spemann really thinking? And what do his thoughts tell us about biology and the study of life?

Reading Spemann's Silliman Lectures makes clear that he was ready, willing, and often able to transplant just about anything experimentally. He had learned so much from transplantation—

from one organism to another, within the same individual, within species, and across species—that it made sense to continue this line of research: taking pieces out to see what difference their absence makes, adding pieces to see what difference that makes, moving pieces around, and exchanging pieces. All this was in the interest of gathering new information about how the developing organism responds. And what is a living organism? Evidently it is a dynamic, responsive whole that can respond to changes and accommodate some and not others. Spemann raised the possibility that some cells retain what Driesch had called "totipotency" much longer than others, and that we might use transplantation exercises to explore this notion. Induction leads cells to differentiate in one way or another.

Induction apparently initiates the onset of processes that lie potentially in the egg. The "organizer" tissue plays a vital role in organizing the organism and the life that results. In Spemann's conception, it made no sense to argue, as even some of his own students began to do, that all manner of material (including inert and killed pieces) might serve as an organizer. They might stimulate a process like induction, yes. But serve as a true organizer? No, for to Spemann "A 'dead organizer' is a contradiction in itself."[25] We do not understand how to interpret the complex results that are coming forth so rapidly, Spemann argued. "When progress is as rapid as it is at present in this domain of science, it is not too difficult to wait patiently, until, by such experiments or analogous ones, step by step a firm footing may be gained." In science, Spemann felt,

> I should like to work like the archeologist who pieces together the fragments of a lovely thing which are alone left to him. As he proceeds, fragment by fragment, he is guided by the conviction that these fragments are parts of

a whole which, however, he does not yet know. He must be enough of an artist to recreate, as it were, the work of the master, but he dare not build according to his own ideas. Above all, he must keep holy the broken edges of the fragments into their proper place and thus ultimately achieve a true restoration of the master's creation. There may be other ways of proceeding, but this is the one I have chosen for myself.[26]

That attempt to hold the pieces "holy" while also exploring possibilities kept embryologists busy transplanting and pursuing other avenues of research. The "fantastical" experiment that would bring together a nucleus from one cell or organism and one from another cell or organism remained elusive for fourteen years. Spemann himself said no more about it, nor did he emphasize the value of what seems today so obviously compelling—namely, testing the relative contributions of nucleus and cytoplasm and the restrictions of totipotency (the potential for a cell to become an entire organism, as the egg cell does) as the embryo develops.

In 1952 Robert Briggs and Thomas J. King did carry out a version of Spemann's experiment.[27] They took nuclei from donor frogs of one species and put them into host frog eggs of another species and also into the same species so as to control as many factors as possible.

Their question was very clear and apparently simple: to test whether the individual cells, blastomeres in the earliest stages of cell division, "remain equivalent or become differentiated as the various parts of the embryo differentiate." Morgan had suggested that nuclei might differentiate in the different cells in response to the environmental conditions of the surrounding cytoplasm. Since the egg cytoplasm is already differentiated in frogs,

WHOSE VIEW OF LIFE?

with areas of "grey crescent" and more- and less-dense parts, this interpretation offered an intriguing explanation for how cells that are all the same genetically nonetheless lead to differentiated cells and parts and ultimately to a fully organized differentiated organism. Therefore, testing whether the cells remain equivalent, and if so for how long, would add considerably to our understanding of how difference arises from sameness. As Briggs and King saw it, "Obviously this problem can be solved only by the development of a method for testing directly whether nuclei of differentiating embryonic cells are or are not themselves differentiated."[28] Transplanting nuclei would make that possible.

When they tried the experiment, they succeeded with nuclei from eggs in the earliest stages of division. That is, nuclei from the blastomeres, resulting from early cell divisions up through the late blastula stage, did proceed to cleavage and full development. They could not, however, get successful transplants using the later stages that Spemann had imagined as so enticing. In most cases, the eggs transplanted with later-stage nuclei would not even cleave once, and they would never divide for long or normally. Briggs and King speculated that it was not possible. They concluded that by the end of the blastula stage and the beginning of gastrulation, the nuclei have changed and lost their capacities or potential to develop the whole.

Since understanding differentiation was precisely the most interesting problem at the intersection of heredity and development, it remained an open question in search of other approaches. Briggs and King saw their results as the very beginning of a line of research and concluded: "These and other experiments prove that the blastula cell nucleus can be transplanted in undamaged condition, indicating that the technique of nuclear transplantation is now sufficiently well developed so that it may be used in studies of nuclear differentiation and possibly in other studies of nuclear function as well."[29]

Briggs and King's transplantation experiments are intriguing in their conception and their possibilities. The idea of transplanting nuclei had made sense, as we have seen, for some time. But how to do it? As a graduate student studying embryology and its history with Briggs in the mid 1970s, I listened as Briggs described these experiments and also his fascination that the public did not get more excited—either positively or negatively—about the possibilities for cloning.

This was the first animal cloning. The term had first appeared in agriculture and botany, when Herbert John Webber wrote in 1903 that "clons" are "groups of plants that are propagated by the use of any form of vegetative parts such as bulbs, tubers, grafts etc., and are simply parts of the same individual."[30] Through the 1920s, researchers had been cloning—in the sense of generating genetic copies of—a variety of asexually reproducing life forms, including tree runners, regenerating worms, bacteria, and cells. The work of Briggs and King was a huge jump forward in terms of workable methods demonstrating what could be done. Yet it fell logically and even methodologically in a set of traditions that pointed in that direction.

The techniques that Briggs showed us graduate students in the lab started with making glass needles. By the 1970s, it was easy to fill out an order form and have a box of excellent instruments delivered right to the laboratory door, of course. Yet Briggs felt that we should make our own, or at least try to, to understand what it was like to discover the possibilities, limits, and challenges of research. There we sat, trying for hours to take thin pieces of glass, hold them over the heat source, and when they got to just the right temperature, pulling quickly and decisively on the two ends. This should produce a neat and very sharp point on each end. When Briggs did it, it worked. We were not so skilled, and we learned very definitely how difficult even this apparently simple step could be. Usually, we pulled too hard or too fast and ended up with a string too thin and fragile to use. Or we pulled too slowly and ended up with a very blunt instrument. Even when we got it almost right, and had just the right point with the thinnest hair of extra glass off the end, we would blunt the needle trying to remove that thin hair.

Research is difficult, of course. Other laboratory tasks were even harder than making the needles. The point is that Spemann, then Briggs and King, then others decided to do them. They had to believe that the method could work and that the details to be learned were worthwhile. In the case of cloning frog eggs, they had to get the needles, get the eggs, and remove the nuclei with either the fine glass needles that Briggs insisted worked best or with micropipettes. This required removing the jelly coating that normally surrounds frog eggs in the early developmental stages. That part is relatively easy, but it requires assuming that the egg does not really need that coat and will develop just fine without it. These steps yielded the host enucleated eggs.

Briggs and King then took nuclei from the donors, and what came next was in some ways the most difficult part. First, they had to conduct many breeding crosses across varieties and species to discover which species are capable of producing viable hybrids, since not many are. Then they had to prepare the host eggs. Pricking them with the sharp needle caused the eggs to rotate and the nucleus to move to the surface, where it was relatively straightforward to remove it. This procedure left the eggs still "alive" and receptive to the new nucleus isolated from a separate egg.

In fact, this separation was not perfect. Briggs and King described selecting one cell from a blastula and dissecting it away from its neighbors: "The cell is now drawn up into the mouth of a thin-walled glass micropipette, the lumen of which is somewhat smaller than the diameter of the cell. The pipette is held in a Leitz-Chambers holder connected via rubber pressure tubing to an ordinary 5-ml syringe. All of the system except the tip of the needle is filled with air. The tip contains the column of solution drawn up with the cell. Provided the needle is really clean the movements of the column can be controlled accurately. Now, as the cell is drawn up into the needle it is compressed and distorted in such a way as to break the cell surface without dispersing the cell contents."[31]

In other words, they did not get the nucleus clean and separated perfectly from the surrounding cell. Rather, some of the egg cytoplasm came along with the nucleus. Along with the donor nucleus, they inserted a bit of extraneous material from the donor cell's cytoplasm into the host egg. They had to be careful not to cause so much disruption that it would kill the egg. Even after all this, some of the frogs developed, and they observed the conditions under which that happened and those under which it did not. What they concluded—besides that it was possible to do this experiment, which held great promise for other studies— was that the nuclei of cells in later stages did not seem able to provoke a series of cleavages and development. Late-stage nuclei seemed to have sustained irreversible changes. Cloning a Dolly, using nuclear transfer from somatic cells, should not be possible according to their results.

Briggs and King had worked primarily with the frog *Rana pipiens*. Across the Atlantic Ocean in England, John Gurdon was working with another frog, *Xenopus laevis*. In some ways, *Xenopus* was a better research subject. As in Briggs and King's work, Gurdon's research demonstrated that transplanted nuclei stimulate development and differention through the tadpole

WHOSE VIEW OF LIFE?

stage. Furthermore, and contrary to the earlier results, Gurdon discovered that while about 30 percent of the nuclei transplanted from blastula stages could produce tadpoles, 6 percent from hatched tadpoles and 3 percent from swimming tadpoles also could. We might say that "only" so few nuclei could be transplanted successfully, but recall that Briggs and King concluded that it was not possible for *any* later-stage nuclei to provoke differentiation, as they were presumed already to have become too differentiated. Gurdon's results reopened the possibilities for cloning beyond just the early stages.[32]

It would be a long time before somatic cell nuclear transfer would be successful in mammals—before Dolly the sheep could be created—but it no longer seemed impossible. It no longer seemed so clear, as Briggs and King had concluded, that the nuclei of those later cells were too far differentiated to promote full differentiation and development. Gradually, in the 1960s, the field of "embryology" as the study of embryos began to give way to the study of "developmental biology" (the study of how genetics informs differentiation and development). The existence of the Society for Developmental Biology, for example, proved that this was a real academic field. Perhaps even if the nuclei were all the same, as Morgan had pointed out, it was nonetheless the case that they are expressed differently in different cells and as development and differentiation progress. It was to that expression, the connection of development to genetics, that researchers began to turn their attention in the 1960s. Their efforts led to cascading success by the end of the twentieth century.

Now, in hindsight, the prospects for cloning and developmental genetics seem obvious. At the time, however, even such enthusiasts for genetics as James Watson, John Tooze, and David Kurtz remained skeptical. In their 1983 textbook, *Recombinant*

DNA: A Short Course, they wrote that "In the immediate future there is little likelihood of nuclear transplantation being attempted with any other mammalian species," or in any other species beyond frogs. "If the efficiency and reproducibility can be improved, the method may, however, find a place in animal breeding. In theory it could be attempted with human eggs and embryonic cells, but for what reason? There is no practical application."[33] History reminds us how difficult it is to predict the future and that we should not foreclose opportunities on the basis of assumptions that so often turn out to be wrong.

Recombinant DNA, IVF, and Abortion Politics

"It may well be that there are some technologies that you should not use, not because they can't work but because of the social dangers involved and the repression that would be necessary to prevent social danger . . . How do you cope with this new observation that some kinds of knowledge and some kinds of technology can be very dangerous? We have no assurances that science will not lead us into a very dangerous world . . . How do you control that without interfering with a lot of the freedoms that scientists have cherished? That is something we are only groping toward . . . How do you make policies for an issue which may take 50 years to resolve? Our government, at least in the past, has not been ready to make long-term decisions."

This was not said about cloning or stem cell research. Instead it was 1976, and Robert Sinsheimer, as chair of the biology division at Caltech, was concerned about the technologies for recombining DNA. As he explained in an interview with Nicholas Wade for *Science,* "Some of my colleagues feel that it is the scientist's job to do science, and society's job to cope with what he does. I disagree with this in principle. The scientist must keep the public informed and involved because nobody else will."[1]

This bold expression of both concern and responsibility was all the more remarkable in coming from one of recombinant DNA's leading researchers, who also was a member of the National Academy of Sciences and editor of the *Proceedings* of that august body. Initially one of the leading advocates of genetic engineering, since the start of the explosion of recombinant research, Sinsheimer had become increasingly troubled by the possible consequences of the research and its biotechnological applications. He had also become less confident of the ability of public bodies to understand and react intelligently and quickly enough in the face of serious public health threats. The news story in *Science* immediately following the interview with Sinsheimer, "NSF: New Program Criticized as 'Appalling' Subsidy to Activists," makes Sinsheimer's point rather nicely. In that report, we learn that in 1976 critics opposing Senator Edward Kennedy's congressionally funded program to promote public understanding of science felt it was a waste of public money. Yet, as we might still point out today, if the public does not understand science very well, and scientists want to pursue knowledge as free inquiry and to let others decide whether and when to use the knowledge, then how will those uneducated "others" be able to make intelligent decisions about science?

Looking at recently developed guidelines to contain research and its potential applications, Sinsheimer felt these did not go far enough. He argued that there are rare cases in which it is appropriate, desirable, and even morally necessary to restrict scientific inquiry. Recombining the DNA stuff of heredity was—at least at the moment—one of those areas. As *Science* reporter Wade pointed out, however, at least for the moment (late 1976) Sinsheimer seemed to be the only one with concerns and "to

WHOSE VIEW OF LIFE?

have a virtual monopoly of long-range thought about the issue." What was really at issue, and what provoked his response?

Recombining DNA

In 1973, discussion about recombinant DNA, as it was called, had gone public. Humans had been recombining DNA for millennia, of course, through agriculture and domestic breeding. By the mid-nineteenth century, Charles Darwin and other architects of evolutionary theory had even begun to understand breeding and hybridization in terms of evolution. By the early twentieth century, they had added the Mendelian-chromosomal interpretation, in which breeding was a matter of genetics and development, all in the context of evolution. By midcentury Briggs and King and then Gurdon had produced a particular sort of hybrid that experimentally brought together inherited genes from different organisms. In fact, genetic recombination was common, in nature as well as the laboratory.

Something had changed by 1973, nonetheless. Scientists had, step by step, discovered how to isolate a restriction enzyme that could cut DNA at specific "restricted" sites. The enzyme acted like a sort of scissors for making specific kinds of cuts. The innovation was acknowledged as significant so quickly that by 1978 Werner Arber, Daniel Nathans, and Hamilton O. Smith had already received the Nobel Prize for Physiology and Medicine for the discovery of restriction enzymes and their application of the techniques of molecular genetics. Others had learned to isolate a ligase that can join together DNA segments.

Cutting apart and putting back together does not sound particularly difficult, nor earth-shattering, but it was. With this technique it was possible to cut one strand of DNA apart and then put it back together with an extra piece or with different pieces. It was possible to "recombine" DNA. Not only that, but it was even possible to recombine DNA from different species. This was spectacular. Perhaps we could have "designer genes" with precisely the combination we want. Add the recently developed ability to isolate and clone genes (that is, make exact replicas of them) so that there was plenty of whatever we want off the shelf to splice in, and this research theoretically would make it possible to generate just the combination of DNA a researcher wanted. In 1973 came the first successful cases of splicing foreign pieces of DNA into circular plasmids of DNA from bacteria. The plasmids were then reinserted into the *E. coli* bacteria. Recombining DNA was not just theoretically possible. It worked.

How exciting! Just imagine the possibilities for helping a human patient who had the misfortune to inherit DNA that was defective in some specific way. Perhaps the capacity to process certain carbohydrates is missing, for example: previously doomed to disease and perhaps early death, this lucky individual might be able to lead a normal life through gene therapy. In theory, scientists could simply find a "healthy" copy of the missing gene, splice it into some bacteria or other carrier or "vector," then introduce it into the patient. People quickly began to imagine applications, such as splicing a gene for insulin production into diabetics. The prospects seemed tremendous, and both scientists and the public quickly embraced the possibilities.

At a meeting at the Massachusetts Institute of Technology in 1973, however, discussions revealed how little researchers really understood about this science and also hinted of the potentially profound problems, limitations, and implications of recombinant DNA technology. If scientists could insert pieces of DNA into plasmids, then put those into bacteria, like *E. coli,* that reside naturally and ubiquitously in human intestines, might the procedure pose serious risks for individual humans contaminated by the introduced DNA? Might not recombined genes escape the laboratory and disturb the environment? What if the apparently innocent pieces of DNA had severe pathological implications for humans? What if oil-eating bacteria would eat not just the "bad" oil from tanker spills but also the "good" oil that we need to fuel our cars and heat our homes? The public press was filled with such scenarios, some raised by scientists themselves.

The specter arose of altering the course of evolution by recombining pieces of DNA that would normally have natural boundaries because they reside in different species. Perhaps the chimeras created through this method would take on a life of their own and become monsters. We knew little about what natural barriers evolution offers to development, but surely there are some. We would be intervening in nature, ripping down walls and protections that we may really need, without having any idea of what to expect. Whether this research would lead to "playing God," acting with good intentions but making mistakes in ignorance, or experimenting intentionally and recklessly, like Dr. Frankenstein, those meeting at MIT in 1973 concluded that scientists should think about what they were doing and act accordingly or be regulated.

Numerous excellent histories have recounted the early debates about recombinant DNA in detail.[2] The stories need not be repeated here, but we should appreciate the importance of this episode that so quickly changed the role of biology in society. With recombinant DNA technology, biology "grew up"; it became a heavily funded, publicly valued science that now dominates the landscape around MIT in Cambridge, Massachusetts, several areas in California, and increasingly many places in between. This work also has challenged what we mean by "a life." If an individual life is defined in terms of inherited genetic material and its expression through embryonic development, is that definition changed when we recombine the DNA? Are we creating a new or different life? If so, does that matter? Are we redefining life or designing life?

In 1973 Maxine Singer, then from the National Cancer Institute, served as co-chair of the summer Gordon Research Conference on nucleic acids. Begun in 1931 and taking on the name Gordon in honor of the long-time administrative contributions of Neil Gordon in 1947, the Gordon Research Conferences have provided a prestigious forum for groups of researchers to come together and discuss the latest scientific work in new and "hot" areas. The rules are clear that these meetings are for exchange of new ideas in an open and supportive environment. Participants sign agreements not to cite or quote others without permission, and the emphasis is on promoting and provoking innovation and reflection. As their website declares, "The Gordon Research Conferences provide an international forum for the presentation and discussion of frontier research in the biological, chemical, and physical sciences, and their related technologies."[3]

Even within this context of predictably exciting science, writer June Goodfield reported that Maxine Singer found the 1973 nucleic acid conference "a very dramatic occasion."[4] Both the prospects and the problems for recombinant DNA work began to emerge, and by the end of the week organizers had decided to dedicate fifteen minutes at the end of a session the last day explicitly to acknowledge the implications of the research. The scientists at this intense research meeting decided that examining social consequences of science is a legitimate topic for discussion even at a scientific meeting—at least for fifteen minutes.

At the end of this brief discussion the participants, realizing that fifteen minutes was not nearly enough time, wrote a letter to the National Academy of Sciences noting the potential problems with recombinant DNA research. This was relatively noncontroversial. The decision to write a letter to the much more widely distributed journal *Science* and thereby to "go public" with the concerns was not so obvious or so popular. The majority agreed, however, and a letter appeared on December 21, 1973.

That letter evoked an immediate public response that, in turn, led to formation of a committee headed by Stanford geneticist Paul Berg and including James Watson, David Baltimore, and others who have remained active both in science and in the public eye ever since. On July 24, 1974, they published in *Science* what came to be called the "Berg Letter." This letter called for delaying research until scientists could assess the potential dangers of work that might put humans at significant risk. Scientists should not splice antibiotic-resistant plasmids into bacteria, for example. In addition, they should be careful with "shotgun" experiments that take all of an organism's DNA and chop it up into random pieces. These experiments seemed to offer the greatest risks: as there is no way to know just what the pieces of DNA code for, there is risk of culturing and combining pieces in unpredictably dangerous ways. The signers of the letter were willing to consider that perhaps DNA splicing should not take place at all. To take practical action, they recommended that the National Institutes of Health set up a committee to evaluate hazards and to develop procedures for containing and controlling the research and its attendant risks.

The letters and calls for reflection led to a conference at the lovely conference facility in Asilomar, California. There scientists discussed what they knew about the science and compared notes. For those who attended, this was not a time for preserving secrecy or protecting proprietary knowledge but for building a solid foundation of knowledge and setting guidelines that could make the world safe for at least some research to continue. For how would we ever learn more if we did not carry out new research? How would we even know what the risks really are? Surely it should be possible to develop national and then international guidelines for containment of potential risks. Soon, sets of protocols for physical and biological containment were put in place with guidelines under NIH purview, and some of the less risky research continued while some experiments involving splicing human genes remained on hold. The Recombinant DNA Advisory Committee (RAC) was established in 1975 with the mandate to provide advice to the NIH director on "conduct and oversight of recombinant DNA work" and, in anticipation of future developments, other activities relating to DNA technology.[5]

Every research campus across the United States that received federal funding set up oversight committees and assessed the costs and benefits of establishing laboratories with the proper levels of physical and biological containment to do this new research. How high would the cost be if genetic material escaped from a high-level containment lab, for example, and how high might the gain be if the researchers made financially successful medical discoveries? Most universities opted for lower levels of containment and sought to reassure the public and campus critics of their watchfulness and wisdom.[6]

Taking these measures now seems a sensible reaction to a new science with unknown potential impacts. Yet, it was so extremely unusual for a group of scientists self-consciously and purposefully to act in this way that people were not sure how to react. A few applauded the scientists for taking the lead in reflecting on the possible uses or misuses of their work. Far more responded in ways that in retrospect look like overreaction. Much of the media and members of the public who pay attention to science were alarmed. If scientists were admitting that their experiments presented some risk, and if even the most expert of experts did not know what the risk might be, then perhaps we should stop such scientific research now and forever. Perhaps there should be limits to what research scientists are allowed to do. Perhaps, as Sinsheimer suggested a bit later, there should be limits on free inquiry for the public good.

The fact that this idea was taken seriously revealed a significant shift in thinking about science and its place in society. In the immediate flush of World War Two success—or what was taken as success by physicists whose atomic bomb was credited, by many, with ending the war and saving countless American lives—science had become correlated in the American mind with progress. Vannevar Bush, advisor to President Eisenhower, had insisted that the United States has an urgent need for publicly funded science. We need basic scientific research, ranging over all scientific fields, because it is on such basic knowledge that medical and technological progress is made. "Advances in science," Bush sought to persuade the president and the Congress, will "bring higher standards of living, will lead to the prevention or cure of diseases, will promote conservation of our limited national resources, and will assure means of defense against aggression. But to achieve these objectives—to secure a high level of employment, to maintain a position of world leadership—the flow of new scientific knowledge must be both continuous and substantial."[7]

There is, indeed, an "endless frontier" that is both challenging and ripe with opportunity for scientists who engage in open, basic scientific inquiry. This means, Bush argues, that scientists should be free to explore science and also that the results will be good for society. Therefore, the reasoning goes, the public should fund exploration for the public good. Young citizens with the necessary talents should step up and do science. In addition, the nation should begin a program to educate more scientists and must fund that educational initiative.

What is the proper role of the scientist? Is he or she a laborer whose work is producing knowledge that is in itself "pure" and "good"? Or is the product at least neutral, so that what matters is how the public, as consumer, chooses to use it? As Samuel Florman wrote about technology in his lovely book *Existential Pleasures of Engineering,* engineers can make motorcycles or oboes, and it is up to society and individuals to choose which they want to use; technology and science do not dictate the society's values.[8] Should scientists be given complete freedom and be expected to police themselves if their efforts go awry? Or is that too much like letting the lunatics run the asylum? What is the appropriate social contract between scientists and society?

Sinsheimer called for at least some scientists to assume a role as public scientists and to take responsibility for reflecting on the implications of research and considering possible restrictions or limits. Knowing that some scientists were assuming such responsibility might even help the public to trust scientists. But should they? And if the public should not trust scientists, and if the public has a right and even a responsibility to impose limits on science for the public good, should scientists trust the public to do so wisely? If scientists were to police themselves, could they trust each other? How do we go about establishing the appropriate balance of rights and responsibilities?

In the mid-1970s, public discussion intensified quickly. While Singer and colleagues felt they were calling for a reasoned "pause" in the research until they could learn more, others interpreted them as having called for a halt on all such work. Scientists and the public in Cambridge, Massachusetts, engaged in lively and sometimes angry debate about how much and what kinds of recombinant DNA research should take place there. What should private universities with large scientific investments be allowed to do, and what restrictions should the city council have the right or responsibility to impose? The University of Michigan considered passing its own ban on recombinant DNA research. Debates carried on in different ways around the country, including in Congress. There, Senator Edward Kennedy introduced legislation to restrict some recombinant DNA research. Regulating science by law in this way could have imposed unique and deeply troublesome constraints on free inquiry, perhaps as a reaction to political influence or transient fears due to ignorance or confusion.[9] Scientists were worried.

In Washington, the National Academy of Sciences held a hearing in March, 1977, as some scientists were seeking an end to restrictions and a loosening of NIH guidelines. Maxine Singer again spoke publicly about this hearing, reporting on the heated debate and litany of what had then come to seem unreasonable and unreasoned concerns. She concluded that "It is difficult to take demands for bans on 'all' recombinant DNA research seriously when they come from those who demand to be heard but do not stay to listen." Clearly disgruntled, she noted: "Those now labeled 'proponents' of the research worked long and hard for prohibiting certain experiments and matching containment requirements to estimated risks to others. The cautious analytical approach is a discouraging tactic against uninformed fear, mysticism, and political opportunism. But it must continue; nothing less than science itself is at stake."[10] We must not let public involvement go too far nor dictate reactions excessively or prematurely, she suggested. Her reflections then and now, on more recent biological issues, raise important questions about the role of expertise in a constitutional democracy. How do we best adjudicate among competing claims in arenas where the area of expertise is itself under scrutiny?

By June 1977, Senator Edward Kennedy had withdrawn his proposed legislation to regulate the research and announced that the scientists had acted responsibly and helpfully. He urged that the NIH guidelines be extended one more year and that public involvement in the discussions about "evaluation, development and implementation of our national policy toward science and medical research" should continue.[11] Philip Abelson, as editor of *Science,* concluded that "Today recombinant DNA research is highly productive, highly competitive. Workers are under temptation to take shortcuts. But they should behave as if their every act is under scrutiny, for indeed it is—by assistants, colleagues, or competitors. A scientist who furnished the pretext for restrictive legislation could count on the ill will of many of those he or she most wants to impress."[12]

This series of reactions is important because it informs the responses that these same scientists, and their heirs, have about new research today. To be seen as expressing any concerns might be taken by the public as calling for a moratorium or for legislative control over science. Especially given the continuing lack of public understanding of science, scientists have little reason to trust that congressmen or regulators will have an enlightened understanding of the science of cloning or stem cell research. Nonetheless, with recombinant DNA there was reason for congressmen and the public to trust the scientists. After all, the scientists did police themselves. They called for a pause when they thought there were questions. They called for guidelines and reflection, and for taking small steps and considering along the way the potential risks and benefits. They helped develop workable guidelines and oversight mechanisms on campuses and in

NIH-funded laboratories. This was a remarkable moment in the history of science.

The public displayed a typical pattern of response, beginning with initial concern followed quickly by an extreme reaction of fear of the unknown and dread of "playing God"—a mixture of worry and anticipation that played out in ways confusing to the public and the eager media. These sentiments quickly gave way to a sense of urgency among the supporters of science to find ways to allow the research to carry forward safely. The promise of the new knowledge might well outweigh the risks, and not recognizing that possibility was itself a risk. So, initial reactions to recombinant DNA research, ranging from acceptance to worry or dismay, were followed by anticipation. Was this a legitimate shaping and redesigning of life? If so, should we seek to reshape and control science?

Watson, Tooze, and Kurtz reflected on this cycle in the introduction to their handbook in 1981. A note in the preface reveals their particular perspective: "The early years following the production of the first recombinant DNA molecules were not marked by the frantic competitive happiness that usually accompanies the opening up of a new scientific era. Instead of dreaming largely about what marvelous new discoveries the next few days, months, or years might bring, scientists were for the most part concerned about whether these discoveries might be prevented from being made for an indefinite period, due to the existence of stringent regulations governing the rise of recombinant DNA. Fortunately, our worst fear proved unfounded, and today most forms of recombinant DNA research are no longer subject to any effective form of regulation."[13] It is not quite clear whether their "worst fear" was about what might happen if the science went wrong or what might have been done to keep them from doing the science they wanted to do—or both. Happily, from their point of view, those "troubled years" of Asilomar and regulation were over. Much better that the public not restrict what scientists can do, they imply.

Indeed, after a few years of restrictions and containment, research did resume. By 1977, Genentech had become the first biotechnology company promoting the prospects for recombinant DNA technology. Even before much research had begun, its stock prices soared and enthusiasm for the project was astonishing to many. Initial concern gave way very quickly to the hopes and dreams of medical miracles. Biotech companies appeared on the covers of magazines, on the evening news, and in stock and business reports. DNA spliced and transferred into mouse embryos that were then implanted in the wombs of female mice produced offspring that showed some expression of the introduced genes. News like this offered considerable promise for gene therapies. Mouse genes could be spliced into other mice; human genes could be spliced into mice; and on and on. Were the new biotechnology firms playing God? The business god and the biotech god seemed to be joining forces to realize the rich promise of this research.

What is life? What is a life? For a while the answer might have been this: the process of becoming an individual life begins with one egg, fertilized, with nuclear chromosomes and genes inherited from both parents and expressed during development in response to the particular environmental conditions. But now that we could recombine the DNA, we could redefine life, recombine life, even redesign life. The arguments that this new capability would incite were just beginning in the turbulent 1970s.

The recombinant DNA debates of the 1970s offer lessons for today. Historian Charles Weiner reflected that his initial impression in documenting the episodes in Cambridge was of scientists' efforts to make the public aware of concerns; then they shifted to becoming defenders and even developers of the science and its applications—apologists for an industry. He concluded, "The challenge now is to involve the entire society in making choices about the uses and abuses of this science and to draw the line to prevent unacceptable applications. In 1971, Leon Kass observed that 'biomedical technology makes possible many things we should never do.' Now that many more things are possible or proposed, including human germ-line intervention for medical or 'enhancement' purposes, we must not let the market decide. We need to be prepared to 'just say no.'"[14]

Carl Feldbaum provides a different view. "Then as now, critics and politicians feared science was going too far too fast, that we were only the seeming masters of technologies that would overwhelm us and our progeny." Yet they stopped to reflect, and moved on. Subsequently, "biotechnology products have helped more than 250 million people through innovative drugs and vaccines." Now Feldbaum, as president of the Biotechnology Industry Organization, urges that we should, in fact, repeat history. "We should take the more sensible step of prohibiting only reproductive cloning—that is, implantation of a cloned embryo"—and should not prohibit other human embryo research. "Twenty-five years ago, when the future of recombinant DNA technology was at stake, hope prevailed over fear, and reasoned debate over sensationalism. We must do our utmost to ensure that history repeats itself in the debate now before us."[15]

Reproductive Biology

It was a short step in the imagination from recombining DNA to designing babies. Yet it was a long way from that mid-nineteenth-century discovery that a life—in the form of organisms, including humans—begins at some point after a material egg is fertilized and initiates differentiation and development. After a century of significant advances in human reproductive biology, women knew a lot more about themselves and their reproductive cycles and birth processes by the 1960s. A combination of the birth control movement, the women's movement, and scientific advances began to transform the mystery of conception and gestation of life.

Though scientists did not readily take up research on contraception, they did study reproductive biology. As leading popular science writer Edwin E. Slosson wrote in 1922 to University of Chicago biologist Frank Lillie asking for a report on Lillie's research, "There is a wide public interest just now in the subject of endocrinology. In fact the public seems . . . to take it up as a fad in succession to the Freudian complexes now going out of fashion." Sex determination, hormones, and embryology were clearly connected. The "heroic age of reproductive endocrinology" followed from 1926 to 1940, with funding flowing and researchers following the "gold rush." As sociologist Adele Clarke has shown, the growth brought the "disciplining" of reproductive biology with programs, publications, and professionalization.[16] And as historian Nicolas Rasmussen argues, the biotechnology industry that arose around hormones in the 1920s and 1930s was not substantially different from biotech development today, except insofar as today researchers can manipulate the chemicals that they study much more than they could then.[17]

Reproductive biology entered agriculture in a big way by the 1930s. Late in that decade, artificial insemination became practical, thanks to the development of sperm extenders. (The "shelf life" of sperm had been a limiting factor to that point.) By the 1950s, experiments on embryo transfer in cows, along with accompanying studies of hormone levels and other chemical and mechanical factors, had become so routine that, as Miles

McCarry noted in 1999, "Almost 60 percent of all calves born on American dairy farms are offspring of parents that have never met." From "it can't work," artificial insemination (AI) and implantation have become standard practice, and "AI is the biggest reason why today's dairy farms are producing at a rate that was unthinkable 50 years ago."[18]

Drawing initially on studies of mammals, such as rabbits or pigs, researchers had over a half-century gradually pieced together the process by which an egg is released from the ovary each month. By the 1940s and 1950s, they understood the human reproductive cycle fairly well through research carried out largely in universities. Earlier in the century, scientists had studied endocrinology and sex hormones, but not reproduction more generally and certainly not birth control. As Norwegian biologist Otto Mohr had insisted in a letter to Julian Huxley, "To me the birth control movement is not a scientific movement, it is primarily a social one."[19]

By the 1960s, enough progress had been made so that leading reproductive biologist Gregory Pincus could summarize the findings on reproduction and birth control, along with the successes of oral contraception, in his textbook *The Control of Fertility*. This was a book full of science, data, and details, published by Academic Press, but clearly intended for the wider audience of scientists, medical providers, and the educated public interested in these important issues. As Pincus began, "This book is primarily concerned with investigations into aspects of reproduction conducted by the author and his colleagues. It is, first of all, an attempt to summarize a collection of data hitherto either partially or not at all presented. Secondly, we shall attempt to indicate those avenues which hold promise for future investigation. Finally, we shall attempt an assessment of the implications of our understandings and ignorances."[20]

He explained, as scientists had known for some time from animal studies, that the ovaries play crucial roles in producing germ cells and also in secreting sex hormones. Since the healthy functioning of ovaries is vital to reproductive health, women and physicians should appreciate the signs and symptoms of unhealthy organs. A balance of the hormone family of estrogen, estradiol, and estrone, along with progesterone and other hormones, stimulates the cycle of reactions that allow an egg, once fertilized by a sperm cell and developed to the blastula stage, to become implanted in the uterus and begin to grow. By the 1930s, scientists had isolated and begun to experiment with synthetic production of estrogenic substances.[21]

Anyone who has seen Woody Allen's 1972 movie *Everything You Always Wanted to Know about Sex but Were Afraid to Ask* will recall Woody dressed in his sperm suit, complete with glasses, neurotically journeying into the female anatomy. Critics then found it tasteless, but by modern standards it is just silly and amusing. The point is that normal, middle-class people learning and talking about sex and reproduction out loud, especially in public places like the movies, was a novel idea. There went the sperm, one among many, on a mysterious journey, duty-bound to try to fertilize the egg. The egg, meanwhile, begins in the ovaries and travels down the fallopian tube to become fertilized, then keeps moving around for a few days until it settles down and eventually becomes implanted in the uterine wall. We knew a great deal about embryos and their development from animal studies. By the mid-twentieth century, biologists had put the pieces together and the public began to express eager interest in the process and patterns of the full human reproductive cycle. Human reproduction was a proper subject for science.

Again, it was accepted that health and vitality require a "normal" balance of sex hormones and activity by the egg, though defining what counts as normal and healthy in the face of incomplete knowledge obviously leads to contested claims and contradictions, some of which remain unresolved even now. As Pincus demonstrated, discovering the intricate mix of embryonic development and hormonal stimulation, as well as the biochemical changes in the uterus generally, required considerable animal research, since human women were not inclined to be cut open during their reproductive cycles—or otherwise. Pincus and his contemporaries presented this animal research as noncontroversial. Of course, they assumed, we will use animals to learn important facts that will help humans live better lives. Animal rights activists, notably led by philosopher Peter Singer, challenged that view and continue to do so.[22]

The need to extrapolate from animals and a few human studies of parts of the process left considerable room for imprecise estimation and judgment. Studies of women's menstrual cycles led to the conclusion early in the twentieth century that there are only a few days of each month when a woman is fertile. That is, during only a few days is the egg in the right condition, the hormones conducive, and therefore the time ripe for fertilization. This was important news indeed, both for couples struggling to have children and for weary women seeking some information to guide birth control. Women were eager for both information and options—whether those options were legal or not.

In addition, researchers began to put together information about normal developmental stages of the egg and embryo and what they knew about women's bodies. Though doctors had accumulated an understanding of reproductive processes during the early decades of the twentieth century, the collation of detailed biochemical and hormonal knowledge increased quickly by the 1960s. In addition, political and social changes increased demand for that knowledge. Women wanted to know. Those women who wanted to help their pregnancies succeed sought out the best medical advice and scientific know-how to ensure the birth of a healthy baby.

Louise Brown

In 1985 Peter Singer and Deane Wells reflected on "the new era" in England:

> On 25 July 1978, in Kershaw's Cottage in Oldham, Lancashire, Louise Brown was born. With her was born a new era in making babies. Until then, every human being had begun her or his existence deep inside a female body. There, unseen by human eyes and protected from any kind of outside interference, egg and sperm had fused and the fertilized egg had begun the process of dividing and growing that leads, if all goes well, to the birth of a baby nine months later.
>
> Louise Brown was different. Not different in her appearance, which was just like any other healthy newborn girl. Nor was this normal appearance in any way deceptive. Beneath the surface, too, there was nothing different about her. Louise Brown was a normal Baby and is now a normal child. It is her history that is different.[23]

Proud parents Lesley and John Brown had wanted a baby, but the eggs that Lesley Brown's ovaries produced encountered a blockage and could not travel down the fallopian tubes to become fertilized and implanted. Surgery did not solve the problem—as it does not in a great majority of the cases of blocked fallopian tubes. This was considered a problem, even though the condition produced no other symptoms and even though some critics later would call into question what they saw as an undesirable "medicalization" of reproduction. The couple wanted to have children, and they could not for biological reasons. To them, theirs was a medical problem, and they sought a medical solution. They went to Dr. Patrick Steptoe, a gynecologist in Oldham.

Steptoe recalled his early years as a medical student, feeling the mother's anguish when the physician told her that she could never have a baby. "Never." He could offer her no hope for a baby, not for her "own" baby anyway. In later years, Steptoe remembered how much he disliked having to say that, but he never imagined then that he would be the one, along with biologist Robert Edwards, to turn "never" into "maybe." This combination of a physician, eager to help infertile women, and a biologist, eager to bring together the science of genetics with embryology and immunology to achieve *in vitro* fertilization, proved immensely successful. Their joint autobiographical account—how they met, how they worked together, and what they did—reveals a close friendship, with mutual respect for each other's specialty and expertise, and an evolving multidisciplinary collaboration.

Though skeptics said it was not possible to fertilize human eggs outside the body, these two believed it was worth trying and continuing to try. After all, infertile women kept trying to have babies, and both men loved babies. First it was necessary to devise a way to "see" inside the woman, to retrieve eggs. Animal studies were easy, especially with inexpensive mice, since the researcher could just kill the animal and take out the eggs. Women, obviously, require a gentler approach.

Fortunately, by the mid 1960s, laparoscopy provided an alternative. Originally used as a diagnostic tool, the tiny instrument made it possible to see and also to retrieve eggs through a small incision. That sounds simple. Before the procedure, however, volunteers had to be given hormones to induce ovulation, a tricky procedure that has unpleasant side effects. This step also poses some risk to the woman's long-term reproductive potential, and it continues to be controversial, especially as some clinics have induced "super-ovulation" to harvest more eggs at once. Removing a larger number of eggs also means that some eggs, if not used as intended, will be wasted, though they could be made available for research.

The next step was to keep the eggs alive in a glass dish and to fertilize them in a way that induces normal cell division. This took many years of searching for what Edwards called the "Magic Culture Medium," basically the medium used for cultivating mouse eggs. Once fertilization succeeded, the question was how to keep the dividing clusters of cells alive long enough to get them ready to transplant back into the mother's womb so that they could attach to the uterus, as normal. There were so many steps to work out, with so many details. Steptoe and Edwards kept trying, and in the face of near-universal skepticism and heated opposition from a few critics on moral grounds, they must have been driven by strong desires to succeed. Fortunately, they achieved just enough periodic reinforcement to keep going: a fertilized egg here, a couple of cell divisions there. Then, growing in a dish, there were four blastocysts (cells after the earliest cell divisions). These gave the two men hope.

In 1976, Lesley and John Brown entered the picture. Steptoe noted that the husband "was devoted to [his wife's] welfare and would not brook high risks to her. She was quietly determined, strong in resolve, unlikely to panic, and would suffer whatever was necessary with stoicism. They were an ideal couple for our attempted treatment." The couple's other attempts had already failed, as they had endured procedures, travel, hormonal upsets, and minor surgical invasions. The physician and scientist explained that their attempt, too, might fail. Yet a "maybe" was better than a "never" for Lesley Brown.

Weeks later, they transferred a fertilized egg and Lesley Brown announced that she "felt pregnant." It turned out that the late-night operation taught them something vitally important. Because they were waiting for the egg to develop to the eight-cell stage before transferring it, the operation did not begin until midnight. Only later did they realize why the procedure had worked at night but not for any of the trials they had conducted during the day: diurnal cycles of hormonal levels proved crucial.[24]

Eighteen days later, pregnancy tests came out positive, and Edwards wrote on December 6, 1977, to Mrs. Brown: "Just a short note to let you know that the early results on your blood and urine samples are very encouraging and indicate that you might be in early pregnancy. So please take things quietly—no skiing, climbing, or anything too strenuous including Xmas shopping!"[25]

Despite the excitement for the Browns and for Steptoe and Edwards, they did not publicly announce the pregnancy nor invite any fanfare. This was, the doctor and biologist decided, a time for patient privacy. To the persistent *New York Post,* which had evidently received a tip from someone in the hospital, Edwards responded that "Any progress in our work will be conveyed to our medical and scientific colleagues through the proper orthodox channels." Unfortunately, the newspaper published that there were "on-going pregnancies," the hospital confirmed the story despite the physician's efforts to get them to wait, and the Browns were thereafter constantly dogged by reporters and had to seek refuge at relatives' homes. False reports, news leaks, and bombardments by the press heightened the stress of an already challenging situation. Steptoe and Edwards did what they could to protect their patient.

Pressures on the Browns must have been immense, and they were aggravated by Mrs. Brown's toxemia and other stresses. The expectant parents had to trust their doctor—their only choice given the pioneering nature of the pregnancy. On July 25, the fifth day of the thirty-ninth week of gestation, Louise Joy Brown was delivered by Caesarean section. John Brown cried, "I can't believe it. I can't believe it!" Lesley Brown whispered, "Thank you for my baby. Thank you." Dr. Steptoe noted that "Louise Joy had arrived, a whole new person to make this family complete at long last. I doubt if I shall ever share such a moment in my life again."[26] It was the wholeness of the baby, and of the family, that struck him. Her birth justified the experimentation with embryos and all the challenges and risks. For Steptoe and Edwards and the Browns, the healthy cries of little Louise completely dissipated the challenges of their critics.

Other babies soon followed Louise Brown and not long after, Australian physicians successfully developed their own procedures. In 1981, the first "test-tube baby" was born in the United States. Very quickly it also became clear that this process of fertilizing eggs in a dish could be useful in cases other than blocked fallopian tubes. Indeed, reproductive specialists soon produced a new sort of hybrid. An egg from one donor and sperm from another implanted into the uterus of a third person could develop into a healthy baby to be raised by other parents entirely.

Though Steptoe and Edwards were persuaded that they were doing good in the world by solving these medical problems for deserving couples, these new "family" arrangements have raised a host of legal challenges. Who owns eggs? Under what conditions is it legitimate to harvest eggs? And who gets to decide how they should be used? It was in 1984 that a woman gave birth to a genetically unrelated baby for the first time, whereas 1989 brought an estimated five thousand egg donations. We read of a woman carrying a child for a sister or of a grandmother giving birth to her own grandchild, and there is news of all manner of "designer" embryos starting with carefully selected donated eggs. As Rebecca Mead pointed out in an article in the *New Yorker,* at first very few women had the opportunity to use donated eggs. The first eggs came from those "extras" left over from the *in vitro* fertilization process. In some instances women undergoing sterilization by having their tubes tied were willing to donate their eggs for the use of others. "Some infertile women were helped by their younger sisters, or by friends," Mead wrote. "There was little concern about matching donors and recipients beyond the broadest categories of race. One recipient I spoke with, who is dark-haired, olive-skinned, and Jewish, received donor eggs ten years ago from a woman who was tall, blond, and Nordic."[27] By 1999, as Mead noted, there was a hot market for human eggs, especially for "desirable" eggs from intelligent, young, attractive women.

Usually when eggs are purchased, a buyer has arranged for the donation and the process proceeds according to the contract. Contract and state law prevails in such cases, given that there has yet been no federal legal ruling to consider eggs as other than commodities owned by the donor. Other laws govern the ownership of sperm. In most cases, sperm are plentiful and not an issue. But in cases of comatose or recently deceased men whose partners want to have their genetic babies, questions arise. By 1999, Lori Andrews explained that sperm retrieval from the comatose and even from the dead had become fairly routine.[28]

Yet who owns *fertilized* eggs? Court rulings have typically ruled that these were also property, though not commodities to be sold just as we cannot sell kidneys or hearts. Yet these eggs sit in a petri dish in a fertilization clinic. Surely, they have a different status there, or in the freezer, than they do once they are implanted into a woman? Indeed, the designated owner can use them, dispose of them, or donate them. As our definitions of what counts as a life shifts, the current patchwork of state court rulings is likely to continue evolving, and increased interstate commerce will eventually lead us to federal legislation regarding fertilized eggs, or at least later-stage embryos.

Once scientists discovered safe ways to freeze fertilized eggs and first succeeded in producing a healthy baby from a frozen embryo in 1984, new questions had to be answered. Who owns frozen embryos, how long do they "keep," and is each one already an individual life, a person alive in suspended animation in the freezer?

Furthermore, as the procedure became more common and the number of fertilized eggs grew, one question became increasingly urgent: what should be done with "extra" fertilized eggs, commonly called embryos? Before successful methods of freezing, the answer was easy: either the embryos were implanted quickly and at particular stages of cell division, or they would die. That's why Lesley Brown had her egg transplanted in the middle of the night; it was ready then. With freezing, though, more and more of the fertilized eggs—"owned" by parents and ruled legally to be "theirs"—survived, stored away in freezers.

What if one parent wanted to destroy them, or what if one parent wanted to "use" the embryos and the other did not? If an embryo was implanted after a parent's death, whose child is it, and what inheritance rights does it have? Courts have made rules in all these cases and many others, calling on ownership and property rights, parental rights and responsibilities, and inheritance rulings for guidance. Yet the diverse court judgments show the complexity of these issues. It seems odd that a child can be judged as a person with rights—including the right to make a claim on parental inheritance even if conceived after that parent's death, as has been ruled in at least one case—but yet that an embryo already fertilized can be considered property and disposed of as such.

We have decided that in the United States women can sell their eggs, which are a valuable commodity. Yet we cannot sell fertilized eggs, embryos, or babies or other body parts. Frozen eggs are sometimes considered property, other times potential persons whose rights and needs should be matters of custody rather than ownership. University of Southern California professor of law and medicine Alexander Capron understated the case in 1999 when he said, "I don't think we've gotten all that far. This whole area continues to be a real headache for everybody."[29]

Initial reactions to Louise Brown's birth were probably more positive because Steptoe and Edwards resisted temptations to rush to the press until the baby was born. Even the invasive press did not hear about the case until the egg had already been implanted and the pregnancy—and waiting—had begun. Critics could not in this case insist that it could not be done, nor could they despair that the process was inherently unsafe and that resulting babies must necessarily be defective in some way. The pregnancy seemed to be such a great medical advance and a cause for hope to so many would-be parents. Enthusiasm far outweighed the concerns and effectively triumphed over the voices of those calling for prohibitions, or at least caution, on moral grounds.

Nonetheless, critics did exist. As Peter Singer and Deane Wells wrote in 1984, "These issues are too important to be left to the scientists and doctors who are creating the breakthroughs."[30] They cited positive Gallup polls immediately following the birth of Louise Brown: 60 percent of those polled favored the technology. Yet the procedures also invoked images of a Brave New World, like Aldous Huxley's, with babies incubated in artificial wombs or designer babies selected from assembly lines, according to the parent's preferences about the "extras."[31]

The chief arguments against IVF, as *in vitro* fertilization quickly came to be called, concerned safety, morality, and cost. Some critics worried about the risks of such "unnatural" procedures. How can we be certain that the process will always produce normal humans? For some, questions remained about whether the process might lead to birth but not to fertility into future generations, until Louise Brown herself gave birth to her own perfectly normal child. A feeling that the "unnatural" was morally distasteful and undercut "proper" and "normal" family values echoed through the conservative media and found expression on university campuses as well. President Bush's selection as chair of his Council on Bioethics, Leon Kass, certainly had held that view, just as he sees cloning, genetic engineering, and many other technological interventions in human reproduction as distortions of the natural process and therefore as "repugnant."

Furthermore, Kass joined others in opposing IVF on the grounds of cost. He became a leading spokesman for the position that such technology is not medicine because infertility is not a disease. The public should not pay for IVF, Kass urged in a campaign that contributed significantly to the current decisions by the government and by most insurance companies not to cover such treatments. Kass argued, for example, that "A conservative estimate might place the cost of a successful pregnancy to be between five and ten thousand dollars. If we use the conservative figure of 500,000 for estimating the number of infertile women with blocked oviducts in the United States whose *only* hope of having children lies in *in vitro* fertilization, we reach a conservative estimated cost of 2.5 to 5 billion dollars." Even if improvements lowered the cost of each procedure, the total would still be staggering. Therefore, he continued, "Is it really even fiscally wise for the Federal Government to start down this road? . . . Much as I sympathize with the plight of infertile couples, I do not believe that they are entitled to the provision of a child at public expense, especially now, especially at this cost, especially for a procedure that also involves so many moral difficulties."[32]

In the United States, IVF or ART (artificial reproductive technology) remains largely unregulated, and only a few states have comprehensive legislation covering these areas. Since medicine is still a state rather than federal matter, and since there is some debate about whether ARTs fall within the proper domain of medicine or within family planning and social services, confusion over competing standards often reigns. Many bioethics and legal groups have advocated comprehensive and thoughtful regulatory legislation to protect patients (or product consumers), and the current excited climate about cloning and stem cell research has increased the intensity of discussion, though not always the wisdom of the deliberations.

With IVF, then, initial enthusiasm was followed by concerns that eventually led to even greater enthusiasm as the business boom encouraged strong support for development. Conservative groups favoring reproductive rights of parents supported the boom. Even Leon Kass has moderated his views, recognizing that IVF has helped many couples otherwise unable to have children.

More recently, as writer Gina Kolata explained in a *New York Times* story, clinics supplying fertility services have expanded so much that they have effectively saturated the market. In the United States, most patients have to pay at least most of their own expenses—and these can be considerable. In the period

1996–1998, the number of IVF procedures increased 37 percent, from 59,000 to 81,000, and the number of clinics rose from 281 to 360. The cost limits available demand, and some clinics have joined the lobbying efforts to require insurance companies to pay, usually on the grounds that all persons have individual reproductive rights to have babies.[33]

As Kolata points out, the market pressures are intense in a climate that is still largely unregulated except by agreement of the clinics themselves and insofar as they follow the guidelines of the American Society for Reproductive Medicine, which has taken the lead in articulating goals and a set of ethical responsibilities. The quest for profits, or at least to avoid losses, can result in bad medicine. "Some experts are deeply concerned," Kolata writes. "With marketing promotions that can shade the truth, referring doctors and patients can find it hard to distinguish between centers with good reputations . . . and ones that promise much more than they can deliver. And some clinics that offer inducements like money-back guarantees may end up doing risky procedures to improve the odds of pregnancy." Clinics may implant too many embryos, for example, or subject women to riskier hormone treatments. Some clinics offer free bonuses or lure physicians with gifts to increase referrals. Others offer discounts and advertise to attract patients. Kolata quotes Stanford doctor Barry Behr: "You have desperate patients. And the potential is there to exploit vulnerable people."

These practices are questionable perhaps, but not illegal or even clearly unethical in our current climate of free enterprise and open business development. Different countries have different guidelines and regulatory mechanisms. Questions about whether and how to regulate the medicine and the business in the absence of clear social or medical policy make it easy to react instinctively rather than reflectively, then to wait until new innovations and applications force another reaction. Expanding medical advances will continue to elicit new social responses and repeated cycles of acceptance, abhorrence, enthusiasm, and tolerance.

Roe v. Wade

Steptoe and Edwards may have helped the Browns with one kind of medical problem, but many more women wanted help of another kind. They wanted fewer babies, or at least to have some control over the timing and frequency of births. Contraception rather than conception. Poor working-class women proved to be especially keen for information and help, perhaps in large part because they did not have as ready access to doctors as wealthier women.

The idea of controlling birth was nothing new. Already by the fifth century B.C., it was regarded as important to declare the proper values of a physician. The guild ethics of the Hippocratic Oath required that a physician not induce abortion. There was already some knowledge of such things. Pessaries provided a mechanical barrier and spermicides a chemical barrier between egg and sperm, and various herbal abortifacients might terminate a pregnancy, though the results were hardly perfect. Folk knowledge accumulated through the trial and error of the empirical approach.

Despite, or perhaps because of, advances in knowledge by the late nineteenth century, "Comstock laws" were passed in the United States to limit information and restrict options. In the 1870s Congress and nearly every state had enacted laws inspired by the ardent anti-vice crusader Anthony Comstock. These laws outlawed dissemination of contraceptive materials and information and defined them as "obscene." Statutes in some states even restricted the information that medical books could present.[34]

Among others, the nurse and social reformer Margaret Sanger found these laws unacceptable and set out to provide contraceptive help despite them. In 1916, Sanger set up the first birth control clinic in the United States, along with her sister, also a nurse, and another associate. In Brooklyn, the three were arrested for distributing "obscene" material, including Sanger's pamphlet, "What Every Girl Should Know." To complement the clinical services programs, in 1923 Sanger started the Birth Control Clinical Research Bureau (BCCRB) as a branch of the American Birth Control League. This bureau lobbied for the establishment of laboratory programs in contraception but had limited success. Scientists with university positions did not want to risk taking up work that their colleagues did not deem respectable, or that was considered more like social activism than "real" science. Nonetheless, advances did come. In 1920, women gained the right to vote, in 1925 Sanger's husband financed the first U.S. manufacturer of diaphragms, and in the 1930s legislative restrictions began to fall.

During the Depression in the 1930s, poverty created even greater incentives to reduce family size and help those women who were lucky enough to have a job be able to keep it. In Franklin Delano Roosevelt's America, sensibilities shifted somewhat and the 1930s saw a coordinated flood of resolutions favoring birth control and increasing pressure to change the old, restrictive laws. For example, in 1936 a U.S. Circuit Court of Appeals judge ruled that Comstock language could not be used to prohibit the importation of contraceptives by a physician for legitimate medical purposes. Since information and contraceptive materials were far more developed and readily available in

Europe and Japan, this ruling was a major advance. Judge Augustus Hand then ruled that new evidence showed the dangers of pregnancy, and he suggested that congressmen would not have declared birth control obscene had they known its medical value in preventing this risky condition. The next year, 1937, the American Medical Association officially recognized birth control as part of medicine. In 1942, the American Birth Control League became the Planned Parenthood Federation of America, and other reformers took up advocacy for the cause. Significant change took several decades, but information and organization moved ahead steadily during the early part of the twentieth century.

All this social and political activity was aimed at striking down restrictions on the current birth control options: herbal or chemical liquids used as douches, diaphragms, condoms, or pessaries used to prevent egg and sperm interaction, and various abortifacients, the majority of them herbal, used to prevent gestation if the egg were fertilized. Research in the 1950s added new options. By then, researchers had acquired considerably greater understanding of the processes involved in reproduction. They thought that preventing the important stages of ovulation, fertilization, or implantation would provide the most likely possibilities for controlling birth. Preventing fertilization was relatively easy with condoms; all that approach called for was manufacturing and increasing access. Preventing implantation appeared most difficult, since neither a mechanical nor a chemical approach seemed likely to be targeted specifically enough. Ovulation looked more promising, especially in light of current knowledge about hormonal cycles.

By 1951, it was clear from animal studies that injections of the steroid hormone progesterone could suppress ovulation, at least under some conditions. The obvious next step was to develop a hormonal contraceptive for women and to ensure its reliability. As historian Elizabeth Watkins explains, most pharmaceutical companies chose to stay away from contraception. Since most states still regulated the sale and advertisement of contraceptives, and since companies did not want to risk the wrath of the Catholic Church, only Searle Pharmaceutical Company accepted the challenge, and it invested modestly at first. Gregory Pincus worked at the Worcester Foundation and served as a consultant with Searle because of his work with cortisone hormones, but he had had limited results. Watkins explains the relationship and why it was generous outside funding from Katherine Dexter McCormick that made possible the research and its ultimate success. McCormick made a series of donations to Planned Parenthood that eventually totaled over two million dollars and were targeted specifically toward developing oral contraceptives.

Pincus continued his chemical research. He sought collaboration with obstetrician-gynecologist John Rock at Harvard for clinical trials. As Watkins explains, Rock brought several considerable strengths. First, as Sanger noted, "Being a good R.C. [Roman Catholic] and as handsome as a god, he can just get away with anything." Second, the Harvard connection lent authority to his work. Third, he did not give the impression of a fanatic; rather, he seemed to have made a reflective choice to support the effort, since "Only in the years after World War II did he become convinced that smaller families, made possible by contraception, were the necessary response to the threat of overpopulation."[35] Rock and Pincus were successful, and in 1960 the Food and Drug Administration (FDA) approved the first oral steroid contraceptive pill. Eight years later it approved intrauterine devices (IUDs) for preventing implantation. Choosing to control the steps leading to the beginning of a human life became an option.

Despite Pope Paul VI's reaffirmation of the Catholic Church's prohibition against "artificial" birth control, the 1960s spawned a sequence of major reproductive changes. In 1969, President Nixon spoke publicly about population and family planning and even called for greater federal funding for family planning services. He linked family planning with larger issues of population control and reflected on one of the tensions in the birth control movement. Yes, greater knowledge and access to contraceptive techniques gave each individual woman greater freedom to control her own reproductive fate; at the same time, however, the idea of coercive use of birth control or propaganda against reproduction in particular populations raised the vision of a return of the eugenic movement. On the other hand, encouraging young women to make individual choices to exercise birth control and offering free or discounted services did serve what was perceived as the public good—economically, medically, and socially. Nixon, perhaps to his own surprise, was a medical and social reformer.

The 1960s and 1970s brought what felt like reproductive freedom for most women. Those of college age need not live in fear of pregnancy if they experimented sexually, and college administrations made decisions about how much to tell students and what services to provide. When Yale first admitted women in 1969, I remember that the health services director immediately held orientation sessions on contraception, venereal diseases, and reproductive responsibility for us newcomers, presumably on the assumption that education is good and making an informed choice is better than facing the consequences of a life-altering error. It was made clear that all of us new women Yalies were expected to act responsibly and not to have to run off to New York for abortions or to drop out of school to have babies. Of course, abstinence was fine, but just in case anybody intended to become sexually active, we all learned the various options. Though we were told of some studies suggesting possible risks and side effects of birth control pills, other studies suggested health benefits from regularized cycles and more "normalized" hormonal doses. The message was clear: science was good; it had provided the means to free women and given us choices.

This was not an easy time, of course. Many institutions and individuals did not take the opportunity to maximize openness and choice. Public school officials often resisted offering any serious sex education, much less information about birth control. Church leaders hesitated to discuss birth control options or sex at all. Education about family planning might be an ideal, but it remained still largely in the hands of physicians or the growing number of trained professionals working in clinics or social action organizations.

Even with contraceptives available, however, mistakes happened. Condoms are not perfect, even when used correctly, and user errors are not uncommon. Oral contraceptives have a very high success rate—but only when taken properly. Since the excessively high dose of steroids in the early versions of the pill made some women feel sick, they occasionally skipped doses or cut back in ways that ruined their effectiveness. Some women could not afford contraception, had no access to contraceptives, or felt moral compunctions about using them. After all, taking decisive action—going to a doctor, getting a prescription, having it filled, and using it regularly—suggested the intention to engage in sexual activity. Some preferred to pretend that sex was an accident and that they could not possibly become pregnant because they were "good girls." Unwanted pregnancies occurred, of course.

Pregnant women had several choices, none good. They could carry the baby to term, with the accompanying risks of pregnancy and the costs thereafter, whether they raised the child themselves or gave it up for adoption. Still today we read of cases of young, middle-class girls aborting fetuses in a high school bathroom, for example, and self-induced abortion was an option then too. Or they could risk a clinical abortion, which in most states was illegal. In either case, abortions could lead to death. Desperate young women bled to death in back-alley abortionists' offices or at home after being punctured by a coat hanger. One estimate in 1990 by the World Health Organization held that 200,000 women still die each year from unsafe abortions worldwide. This may not sound like such a terrible statistic, given the estimated 50 million abortions performed, but when half of those are illegal, which cause the majority of deaths, it is a tragedy. Furthermore, for every death there are another twenty to thirty infections, severe injuries, or permanent infertility.[36] The situation was certainly no better in the 1960s and 1970s. Pressures mounted to legalize access to choices, including abortion.

Socially, the 1960s and 1970s were marked by a strong endorsement of individual civil rights and autonomy. The American civil rights movement focused on racial inequities, of course, and the women's movement addressed a suite of complaints. Calls for reproductive autonomy joined the chorus. In 1971, Congress formally repealed the remaining Comstock laws. Furthermore, with the enactment of Title X of the Public Health Service Act, the U.S. Department of Health and Human Services began funding voluntary programs in family planning. The National Family Planning and Reproductive Health Association was formed, and philanthropic organizations began to support clinics, education, and contraceptive programs.

Most dramatically, on January 22, 1973, the case of *Roe v. Wade* was decided by the United States Supreme Court. In writing the Court's decision, Justice Blackmun recognized that this was a difficult case with widely different moral views informing the decision. He quoted Oliver Wendell Holmes: the Constitution "is made for people of fundamentally differing views, and the accident of our finding certain opinions natural and familiar or novel and even shocking ought not to conclude our judgment upon the question whether statutes embodying them conflict with the Constitution of the United States."

It took some courage on the part of the Court to look beyond existing sensibilities, but times were changing. We need not review this entire case, which has been dissected from many angles, but certain features are centrally relevant to our discussion. First, the decision reviewed the history of abortion and reasons to prohibit it. It was really only in the latter half of the nineteenth century that states banned abortion, the court pointed out. The primary argument for imposing the State's interests over the interests of the mother would be, first, to discourage illicit sex. Yet that was not the case made by the state of Texas in *Roe*. In addition, though not expressed in the decision in such terms, the ready availability of contraception had already seriously undermined that argument. Second, if a medical procedure is risky, the State might restrict its use to protect citizens. That might once have been the case, but the Court's opinion explained that such conditions had changed.

Third, there might be a State's interest, or even duty, to protect prenatal life. If the fetus is indeed "a person," then it should fall under the Fourteenth Amendment protections of the U.S. Constitution. Even if it is only a "potential person," the State would have a legitimate reason at least to be very careful about violating its apparent rights. It is this line of reasoning in the decision that has provoked close attention and debate. The Court ruled in 1973 that the definition of "person" "as used in the Fourteenth Amendment does not include the unborn." This conclusion does not, however, require a definition of "life." A fertilized egg, a blastula, an embryo, a fetus, an unborn child might well be alive but it does not, in the Court's ruling, qualify as a "person."

The decision explained that under the circumstances, "We need not resolve the difficult question of when life begins. When those trained in the respective disciplines of medicine, philosophy, and theology are unable to arrive at any consensus, the judiciary, at this point in the development of man's knowledge, is not in a position to speculate as to the answer."[37] Teachings from the past—from Stoic and Jewish philosophers, some Protestant and even some Catholic writers—allowed that life begins after conception and some time before birth, perhaps at forty days or even later. For legal purposes of this ruling on abortion rights, the Court decided that the Justices need not decide—and indeed could not legitimately decide—when life begins.

In section X of the Court's decision, the opinion noted that women in the first trimester, or first three months, of pregnancy have a greater mortality than women who are not pregnant. Therefore, it is clearly legitimate to terminate the pregnancy during this time, because doing so will be a medical advantage to the mother, statistically. By the end of the second trimester, the fetus has not yet reached independent viability, and therefore abortion may be performed during this time. But "for the stage subsequent to viability," or during the last three months, the State may regulate and even proscribe abortion. Few states did immediately, but many have sought to prohibit what they call "partial birth abortion" and to regulate abortions to the full extent and in every way that the Court rulings allow. They have pushed to test the limits of the ruling.

If the 1960s and 1970s brought such medical advances and social progress toward autonomy and focus on individual rights, whence this resistance? Other developments occurring at the same time brought forth an opposition to abortion. While the women's movement applauded the new contraceptive options and access to legal and medically safer abortion, in 1973 the National Conference of Catholic Bishops was organizing a National Right to Life Committee. Basic access to information and alternative courses of action might seem an obvious goal in a democratic society, but not for those fundamentally committed to holding their particular values as absolute and immutable, without possibility of compromise or accommodation. The clashing positions allowed little room for tolerance. Traditional Catholics found common cause with conservative fundamentalist Protestants and others who saw abortion and even birth control as a form of murder. The foes of abortion came from a small minority of the U.S. public in 1973, but their side has grown since and has gained considerable financial backing. In addition, this group has become more vocal and even violent, at times militaristic and occasionally even resorting to murder in the name of saving lives. Polls continually show that the most extreme pro-life movement represents only a minority of Americans, but many of those have a passionate commitment.

For obvious reasons, this group opposes innovations like the "abortion pill," or the "unpregnancy pill," as RU-486 is commonly known. This is essentially an anti-hormone that acts in a way that its French developer, Etienne-Emile Baulieu, likens to jamming a radio signal. RU-486 neutralizes the progesterone that is necessary for the blastula to implant in the uterus and thereby induces a sort of early abortion, before implantation. Baulieu reminds his readers that more than half of fertilized eggs (or zygotes) "abort" spontaneously and for natural reasons never even become implanted. Genetic abnormalities, mechanical, chemical, and hormonal imperfections: any of these can prevent the zygotes from reaching that next stage. As Baulieu notes, "Any abortion represents a failure—an attempt to reverse a condition, correct a mistake, erase a reality." He continues, quoting Australian reproductive specialist Roger Short, "abortion is like poverty: no one likes it, but it will always be with us."[38]

The interesting debate is not between extremists who wantonly encourage abortions and those who seek to deny them at all. Abortions, as Baulieu points out, occur in large numbers. The important social and medical questions are when, why, and under what conditions they should be allowed to occur. When, and under what conditions, is it acceptable for us to stop what either could potentially become a life or what is already a life in this way? Given the diversity of competing claims, how do we decide? Similarly for birth control: when and under what conditions is it acceptable and even desirable to prevent births? In the case of *in vitro* fertilization, when, under what conditions, and why is this artificial intervention in reproduction appropriate? The questions evolve, and so do the answers, against a background of constant concerns about women having healthy babies.

These questions all concern individual reproductive choices. What happens when we add up all those individual choices so that they make a social impact? That is what eugenicists have always worried about. Again, the questions evolve, but again with the same basic concerns. Gail Collins reminded us in a recent essay in the *New York Times* that we have not moved all that far from the concerns of early-twentieth-century eugenics. She notes that Teddy Roosevelt "thundered" that "If Americans of the old stock lead lives of celibate selfishness . . . or if the married are afflicted by that base fear of living which . . . forbids them to have more than one or two children, disaster awaits the nation." Roosevelt had six children. "G. Stanley Hall," she continued, "warned that 'if women do not improve,'" in producing more of the right sort of offspring, then the men might have to resort to "a new rape of the Sabines." As Collins noted: "It's always comforting in a time of crisis to note that we have been down this road before and are still around to worry about the state of the pavement." She concluded that "the fact that women who choose hard-charging careers often do not have children is pretty far down on the list of American social problems. Anyway, things are bound to improve by the turn of the 22nd century."[39]

Perhaps so. Yet while some promote the social responsibility to have children—for the right people to have the right children at least—others claim that the socially responsible thing to do is not to have children. The latter appeal to the "tragedy of the commons" argument. We already have too many people, too few resources to support them, too little quality of life. Medical ethicist Joseph Fletcher has argued that "reproductive rights are not absolute and those who are at risk for passing on clearly identifiable, severely deleterious genes and debilitating genetic disease should not be allowed to exercise their reproductive prerogative." Furthermore, "testes and ovaries are communal by nature, and ethically regarded they should be rationally controlled in the social interest."[40] Or as Bentley Glass put it in 1970, "in an overpopulated world, it can no longer be affirmed that the right of the man and woman to reproduce as they see fit is inviolate."[41]

Having to share the earth's limited resources, or "commons," may persuade some people to propose that we ought not to have "defective" children, not to have children at all, or to have more of the "right" children. To control births or not to control births. To increase fertility or not to increase fertility. Many people espouse their "truths" and often with vehemence. We must learn how to avoid the tragedy of the commons and how to adjudicate effectively among competing views that are often presented in black and white as incompatible absolutist claims of moral certainty. Without thoughtful scientific and medical policy to guide our actions, furthermore, and without thoughtful reflection on parallel cases from the past, we will continue repeating cycles of reaction and overreaction.

The same kinds of decisions that President Nixon and the nation faced in the 1960s are reappearing, provoked by new developments presenting us with economic and social challenges. Michelle Andrews explains in a recent business commentary that the new choices are likely to be too expensive for many. When Nixon insisted that "no American woman should be denied access to family planning assistance because of her economic condition," he realized that providing public funds through the Title X Family Planning Program would actually save money. President Bush, Andrews suggests, does not seem to realize that every public dollar spent on family planning services saves three dollars in Medicare spending for pregnancy and the care of newborns.

The weekly skin patch and a three-week vaginal ring for contraception are likely to cost more than the public clinics or poor women can afford. We cannot afford *not* to have the full range of options available merely because legislators and the president feel that we cannot pay for such services with public funds. As the executive director of the Family Planning Council (which oversees many publicly funded clinics) laments, "You have to make horrible choices . . . It has always been our practice to provide all birth control methods to all women, regardless of income. But I don't think I can do that anymore."[42] If we decide not to invest publicly to provide choices, then we must live with the consequences—whatever they are, including the consequences of defining, choosing, and aborting what some consider to be fully meaningful lives. Once again, strong arguments are urging us to develop wise social policy that is informed but not dictated by science alone.

5

From Genetics to Genomania

Craig Ventner and Francis Collins were both featured speakers at the AAAS meeting in San Francisco in February 2001. That week's issues of *Science* and *Nature* were there too. Playing the starring role, however, was the Human Genome. The American Association for the Advancement of Science holds its annual meeting over the Presidents' Day Weekend, each time in a different part of the country and meeting in conjunction with the National Association of Science Writers. This guarantees that there are always hundreds of reporters and science writers at the meeting to hear about the latest science. Since the AAAS is the world's largest general science organization, serving as an umbrella for thousands of individual scientists and hundreds of scientific groups, it brings together a diverse collection of people all interested in different aspects of science and its social implications. The Program Office, for many years headed by Michael Strauss, who is himself a trained scientist and a widely read and interesting man, worked hard to make this the meeting of record for all scientists and to offer the most compelling scientific program possible every year.

The 2001 meeting was compelling indeed. This was the week that Ventner and Collins had agreed—working, respectively, with the editors of *Science* and *Nature*—to publish their "draft" of the human genome and to present the important findings and their implications. Strauss and the AAAS had given Ventner and Collins prime places on the program, and everybody at the meeting felt the buzz of excitement and anticipation. As one of my undergraduates put it, "The energy at the meeting was amazing. I couldn't believe that these important scientists were right there in the same room with me talking about the human genome. Just like that." Another, who helped as a session aide in one of the genome sessions, noted that "Dr. Ventner certainly seemed like a self-centered guy who isn't very interested in sharing credit. If that's what you have to do to succeed in biotechnology, I don't want to do it."

Eagerness to see the results published and to hear the lead protagonists pronounce their conclusions had been building for over a decade, but especially during the previous half-year. On Monday, June 26, 2000, the two scientists had been together at the White House. There in the East Room, with British Prime Minister Tony Blair by teleconference from London, President Bill Clinton announced that the race to decode the human genome was complete—or nearly so. Sequencing the genome involved, the president noted, "learning the language in which God created life." The information held momentous potential to have "real impact on our lives—and even more, on the lives of our children. It will revolutionize the diagnosis, prevention, and treatment of most, if not all, human diseases . . . In fact, it is now conceivable that our children's children will know the term 'cancer' only as a constellation of stars." Invoking Meriwether Lewis's map of the great American western frontier, which Thomas Jefferson had rolled out in that same room, Clinton said the new map of the genome was "of even greater significance." "Without doubt," the president enthused, "this is the most important, most wondrous map ever produced by mankind."[1]

Surrounded by leading scientists and government officials, Ventner and Collins shook hands and looked to the world like collaborators on a project that would be great for mankind. Nobody that day in June contradicted the president's exaggerated claims. Even scientists who knew better basked in the glow of cooperation, hands across the podium and hands across the sea. The British were funded largely by the Wellcome Trust, the Americans by Congress through the National Institutes of Health and by private venture capital invested in Ventner's company, Celera. The race had been won—by everybody: by the American people, the British people, the politicians who had had the foresight to provide the funding and start the project, by the future and our children's children, and on and on.

Everybody won and everyone shall have prizes, as Alice had discovered when she fell down the rabbit hole into Wonderland. After the animals and Alice had run around and around to get dry after swimming and then stopped, the dodo announced that the race was over. When the runners asked, "but who has won?" the answer came that "Everyone has all won, and all must have prizes." Only then did Alice discover that she was to provide the prizes. She reached into her pocket and discovered a box of comfits that she handed around.[2]

Like Alice, the president and the public soon discovered that they, too, were expected to provide the prizes that others were so eager to offer. Questions about who should pay for science return with new twists when we learn of the significant profits to be made by individuals who own rights to some of the knowledge thus far publicly supported. It is important to appreciate how this story has developed, as a case of science intertwined in complex ways with society and with society's expectations, in the face of inadequate science policy, growing corporate interests in biotechnology for profit, widespread scientific illiteracy, and competing claims to knowledge and expertise.

Just what was in those *Science* and *Nature* publications in February 2001? What is the Human Genome Project really about, and how did we come to invest so many dollars in it? Was investment in this "big biology" project, often represented by its advocates as a search for the Holy Grail, a "good thing," and who decides, on what grounds? And what are the implications of the resulting "map"? Human genes, even our entire genome, are so little different from those of other species. How did genomics come to have such a gigantic public presence?

Before Genomics

Answering these questions takes us back to hereditarian thinking and the rise of genetics, back to 1953 and the romanticized story of the "race" to discover the double-helix structure of DNA molecules. Before that time, scientists knew that eggs and sperm inherit some Mendelian-type material particles from their parents, and as Theodor Boveri and Walter Sutton had shown those particles are arranged along chromosomes. The Mendelian-chromosomal interpretation of heredity allowed Morgan and his laboratory full of fly "lords" to establish correlations between particular genes and specific chromosomes. This is somewhat easier—though certainly not easy—with fruit flies, which have only four rather large chromosomes; humans, in contrast, have twenty-three.

The 1950s brought new thinking, not only because of the double helix. In addition, researchers began to look in earnest for genes that could be correlated with particular diseases. James V.

Neel, a geneticist with an M.D. degree and medical interests, moved from research in *Drosophila* genetics to human genetics. As historian Daniel Kevles discovered from interviewing Neel in later years, Neel explained: "When I came into human genetics, I had one, I guess absolute, guiding principle: Try to be as rigorous as I would have been had I remained with *Drosophila*. That meant picking problems carefully, problems where we could get solid scientific evidence about inheritance in man." That conviction led him to human blood because it was reasonably amenable to manipulation and experiment: "You can spread it, you can look at it, you can treat it objectively."[3] Neel's studies, combined with others, such as Linus Pauling's, demonstrated that sickle-cell anemia resulted from a particular change in one particular sequence of DNA, or one particular gene, and that the condition resulted from a recessive allele rather than a dominant one, as had been thought. Pauling and colleagues at Caltech showed that the gene allowed hemoglobin molecules to cause the red blood cells to "sickle." And this sickling is what causes symptoms ranging from minor to severe and debilitating. This chain of discoveries about genes and blood inaugurated an enthusiasm in the 1950s to search for single-gene defects, with the hope that scientists would be able not only to identify them but eventually also to correct them or at least to block their effects.

Others looked at whole chromosomes. The term *genome* was used on occasion by the 1930s to indicate the sum total of all the genetic material in all the chromosomes, but little was known about those chromosomes. It was known that the difference between having a Y and an X chromosome or two X chromosomes effectively determines whether a child will be male or female, as long as we ignore the indeterminate cases that might suggest some sort of intersex possibility.[4] Therefore, whole chromosomes might contain packets of other information important for other characteristics. The discovery in 1959 that what was called Down or Down's syndrome, was caused by the extra copy of this particular chromosome, or trisomy 21, provided the first example. What, researchers had been asking with increasing urgency, is the substance and structure of these strands of inherited genes? They had been working away to determine the stuff and structure of DNA, or deoxyribonucleic acid.

How are the ribose, or sugar, and the nucleic acids arranged? As James Watson tells the story in his vastly popular, and to many critics extremely annoying and distorting, autobiographical reconstruction, *The Double Helix,* the race was on. Watson presents the search as a race to riches—not financial riches, though these followed in many subtle ways, but the riches and prestige of scientific knowledge and the satisfaction of winning the chase. He presents it as a fair race. Others involved denied that they even knew they were engaged in a race, however, or claimed that if there was one, they saw it as hardly fair.[5] What matters for our purposes is not the details of that story, but that there was such a story. The fact that Watson presented that episode in the history of science as a race is telling. It undoubtedly influenced his ability to construe the sequencing of the human genome, about fifty years later, as another race.

With the structure of DNA made known in 1953, researchers realized that in theory at least they could begin to unpack the tangle of inherited genes. The job involved translating the sequence of nucleotides that are lined up along the DNA's helical backbone. In fact, the DNA in the chromosomes consists of two strands that are "anti-parallel": one runs in one direction and the other in its reverse. This is because of the beautiful matching of the base pairs in each strand. The adenine, guanine, cytosine, and thymine (A, G, C, T) each have a "mate" on the opposite strand. An A on one strand (under normal conditions) is paired with a T on the other strand, and G with C. The alphabet of DNA is limited. Yet because those few little letters are so predictably organized, they hold tremendous importance for life and for the life scientists who want to understand and ultimately to manipulate them. Of course, life still arrives in individual packages, which still pass through their embryonic stages and turn into organisms like us. But the focus now was on the DNA.

It was as if the focal distance of our scientific lens had changed. Instead of whole organisms, living persons and animals, and developing functioning embryos, the focus shifted to a different level. Imagine looking at something in the sky. First you see one thing, then your eyes shift and suddenly you are looking at another distance and seeing something new altogether. First it is an airplane relatively nearby, and then you see a shooting star in the distance, or shift again and you see the nuts and bolts outside the plane. Or look at the surface of the water in a pond and watch the leaves float by, then suddenly become aware of the fish swimming underneath. In biology, it was as if researchers suddenly saw with greater clarity the DNA that makes up the genes and the chromosomes deep inside. For many, the organism and its embryo faded away.

Looking Inside at DNA

Again, however, the question arose: how to see inside? Again, researchers had to discover ways to trick the organism into letting them see. The earlier techniques for recombining DNA enabled scientists to use a "shotgun" approach and "shoot" out the DNA into a cluster of little pieces (achieved by using restriction enzymes that cut only in particular places). Each piece could then be cloned (copied exactly) again and again. That technique allowed what would have been a tiny, single fragment of DNA—something very, very difficult to see even with the best microscopic equipment—to be multiplied so much that a whole mass of its copies would cover a petri dish. This is the same idea as culturing bacteria or growing yeast for bread or brewing: you begin with a little bit of starter and then cultivate it so that you get more and more. Soon you have enough to see what you are making. Better yet if you can figure a way to culture a large number of sequences at the same time.

Early sequencing techniques involved taking these separate pieces and reading off the alphabetical sequence, or the sequence of nucleotide bases (A, C, T, or G) that make them up. Recording the order of the letters, over and over, for all the pieces in effect made them up into words. The technique produced sequences of short pieces, but the goal was to get at the longer pieces and eventually at the whole. That meant putting the words back into the right order, into sentences and then paragraphs—putting them in the order in which they occurred on the chromosome. Next, the sequences would have to be "mapped" to the right chromosome. That became the goal, but at first the effort of simply capturing the sequences of shorter pieces and then discovering how to put them back together was challenge enough.

The scientists were, in effect, blowing the DNA apart and studying its fragments. Funny idea, really, that taking it apart and doing such unnatural things to it is the best way to learn more about the basic, core substance of life. This molecular material was a long way from the living organisms from which the DNA had come in the first place. Yet as we learned from the history of analytical experimental embryology, the best path to understanding life sometimes begins with the death of the whole and a study of its parts. Reducing organisms to molecules can give us knowledge, of a sort. We will return to questions about what strengths and weaknesses the approach may have for generating robust scientific knowledge about life, the life of an individual living organism.

The rush of scientists into nucleotide sequencing produced a store of knowledge about various bits of DNA. After a tedious process of elimination and collation, molecular biologists were able to establish that some sequences of DNA are associated with some "genes." We believe that these are indeed genes because they are associated with some function, notably the ability to produce a protein molecule. Those are easy sentences to write down, and the concepts sound simple.

In fact, however, what counts as a gene and what counts as evidence that a particular sequence of DNA is really a gene is highly controversial. Establishing which are the genes and which sequences are "junk" or "nonsense" DNA remains a problem whose solution is extremely important for the longer run. If we miss important functional parts of the genome because we consider it unimportant, we may either make mistakes in interpreting the sequence or end up having to go back and do the sequencing over again. This is the reason that the NIH project has emphasized getting as complete as possible a sequence and mapping of all the genome, not just the bits that seem to some researchers to be the important ones.

Many philosophers of science as well as biologists have admitted that even those who are most confident about the importance of genetics disagree about what they mean by a "gene." Typically, the word may have a molecular or chemical meaning, as a particular string of nucleotide bases. Or it may have a structural or physical meaning, since the string has until recently been assumed to reside at a particular place, or locus, on the chromosomal DNA strand. Finally, a gene also has a functional meaning, as coding for the production of a specific protein.

The idea that particular molecular structures, found in a particular place on chromosomes inherited from parents, give rise to particular proteins helped generate the enthusiasm for genetics. In biology, at least, researchers could begin to see how to break down or reduce the complexities of life into units that were easier to study. It is no accident that many of the first molecular biologists moved into biology from the physical sciences. And it is no accident that many molecular biologists today, including probably most sequencers, come from highly technical training programs, many of which do not teach much of what was traditionally considered biology at all. Who needs organisms and interactive organic systems when we can make such progress by focusing on the genes and life's molecular structure?

Naturally enough, the success of sequencing techniques quickly and all too easily led to a drive to discover "the gene for . . . whatever." Morgan had worried as long ago as 1909 that biologists were rushing to turn the observable features of cells and chromosomes into "factors" that were assumed to carry information from one generation to the next. He had seen this jump to interpretation as speculative and unwarranted by the data at hand. The ninety or so years that followed did, in fact, add to those original facts a great deal more raw data and then transform some into factors. By the 1990s researchers sought to turn the factors into characteristics, once again often jumping speculatively far beyond the facts at hand. At first this interest remained positive but subdued. As long as the research remained so difficult and slow, and as long as the commercial possibilities remained unclear, interest remained reserved. That changed, of course, as genetics gave way to genomics.

Genomics and Genomicism

By the 1980s, it was possible to sequence longer strands of DNA. Molecular biologists wanted to study not just separate individual pieces of DNA and not just individual chromosomes, but entire genomes. According to that wonderful record of our use of words, the *Oxford English Dictionary,* in 1987 V. A. McKusick and F. H. Ruddle brought the word *genomics* into common usage, even entitling a jointly authored book by that name. They thanked T. H. Roderick of the Jackson Laboratory in Bar Harbor, Maine, for suggesting the term, which they defined as "the newly developing discipline of mapping/sequencing (including analysis of the information)."[6]

This definition—that not only is there a composite of all the genetic material that makes up the genome, but that it is a field of study—is striking. It implies that there are distinct questions to be asked about genetic material and particular ways to study genomes. The idea was catching, and the possibilities for mapping and sequencing, for "doing genomics," quickly captured the attention of biologists.

By the mid-1980s, genomics was off and running. Technical advances made this happen, but another reason was the vision of big thinkers who helped biologists begin to imagine what was possible. Marvin Carruthers at the University of Colorado developed ways to synthesize fragments of DNA, which allowed additional research, and he and Leroy Hood at Caltech developed equipment for carrying out these syntheses mechanically. Besides speeding up the sequencing process, their technique opened the door to the invention of other mechanical tools for sequencing and identifying DNA fragments. For example, gel electrophoresis, which causes different types of DNA to move along a column at different rates and for different distances, made it possible to sort longer fragments from shorter and to do so faster. Techniques for using fragments of ribonucleic acid (RNA)—the material that makes it possible for the genes made up of DNA to produce proteins—vastly improved the targeting of genes. Among all that vast amount of DNA, if we could pick out the small fraction that actually consists of functional genes, we would tremendously cut down the work of decoding. Craig Ventner realized this and began to use libraries of cDNA, or complementary DNA generated from messenger RNA, which was already known to correspond to the transcribed genetic parts of the whole chromosome. Through the 1990s, he and his collaborators also developed techniques and equipment for sequencing large numbers of cDNA pieces at once, as well as computational systems for analyzing the data. His approach was different from that of the NIH research program and complemented their more cautious research strategies. This rush of technical advances created opportunities for a rush of discoveries.

Paths to the Genome Project

Historian of biology John Beatty provides a valuable overview of the genome project and its relations to perceived national security issues. As Beatty points out, both a longer and a shorter history of the genome project may be told, the difference being what biologists refer to as the ultimate or the proximate causation. The longer, or ultimate, story includes a look at the Department of Energy's (DOE) Atomic Bomb Casualty Commission and its tradition of research in human genetics. James Neel and other human geneticists who played roles in this research saw bi-

WHOSE VIEW OF LIFE?

ological study as important for national security. In addition, others found it useful to package their research in terms of national security.

This longer story takes us to a meeting of experts in Alta, Utah, in December 1984, sponsored by the DOE to discuss studies of the effects of the atomic bombs in Hiroshima and Nagasaki. The group concluded that the DOE needed to promote the development of sequencing technology for genetics in order to be able to assess and interpret the data they had available correlating genetic effects of environmental changes. This goal established an interest that the department sustained even as the original rationale faded in the face of technical advances and shifting research missions of the DOE and national labs. Nonetheless, the stimulus and resulting strong political support proved critical at several junctures as discussion of sequencing the human genome grew into a large-scale project evoking national pride and calling for public investment.[7]

Meanwhile, Robert Sinsheimer, who had raised concerns during the recombinant DNA debates about the proper reach and scope of science, had become chancellor at the University of California at Santa Cruz. As with most university leaders, on learning about the plans to sequence the human genome Sinsheimer quickly turned his mind to fundraising. As a biologist, he thought of big projects and their possibilities. What about establishing a center to sequence the human genome on the UCSC campus? Late in 1984, Sinsheimer proposed an Institute to Sequence the Human Genome to the university president, David Gardner. Gardner did not fund the project, but Sinsheimer nonetheless organized a conference for May 1985 for molecular biologists to discuss the idea.

Apparently independently, Renato Dulbecco at the Monoclonal Antibody Laboratory of the Armand Hammer Cancer

Center of the Salk Institute also imagined a human genome sequencing project. In a May 7, 1986, editorial in *Science,* Dulbecco wrote that determining the sequence would be "a turning point" for the study of cancer and tumor virology. "If we wish to learn more about cancer," he urged, "we must now concentrate on the cellular genome." In addition, if our goal is to understand the role of genes and the genome in humans, we must study human genetics "because the genetic control of cancer seems to be different in different species."

This study cannot be taken on by any single group, Dulbecco noted, but must be a national undertaking. "Its significance would," he claimed, "be comparable to that of the effort that led to the conquest of space, and it should be carried out with the same spirit." Better yet, make it an international collaboration, "because the sequence of the human DNA is the reality of our species, and everything that happens in the world depends on those sequences." Dulbecco saw the promise of a cure in having a complete map of the human genome: "The next generation can look forward to exciting new tasks that may lead to a completion of our knowledge about cancer, closing one of the most challenging chapters in biological research."[8]

At about the same time, mathematical biologist Charles DeLisi was at Los Alamos. He had worked at the NIH and in 1986 was director of the Department of Energy's Office of Health and Environmental Research, which conducted studies of the genetic effects of radiation from the atomic bombs tested and dropped during World War Two. DeLisi saw the potential for a sequencing project at the New Mexico national laboratory. To explore that possibility, he organized a workshop for March 1986 in Santa Fe, where "the excitement was palpable."[9] Some biologists attended both meetings and began to shape what became the Human Genome Initiative. As Walter Gilbert put it at that meeting, sequencing the human genome would be the "grail of human genetics" and would provide "an incomparable tool for the investigation of every aspect of human function."[10]

Clearly things were not just heating up but actually reaching a boiling point in the genome world, judging from the number of meetings and discussions all bubbling up about the same time. At the annual Cold Spring Harbor Laboratory Symposium, held in late May to early June 1986, the topic was the molecular biology of *Homo sapiens*. As Roger Lewin reported, "In effect, the symposium was a celebration of the impact of molecular biology on the understanding of the human condition, including genetic disease, cancer, and evolution."[11]

The Cold Spring Harbor Symposium always attracts attention, and the intense discussions bring together the top biologists across a range of disciplines. The discussion was lively. Unlike the earlier meetings, which had been oriented more toward thinking about how to carry out a genome sequencing project, this meeting featured both questions and concerns. Some biologists, like Maxine Singer of the National Cancer Center, thought that there were better ways to study the genome that would include more biochemistry and more multidisciplinary approaches generally. Others, like David Botstein of MIT, worried that the project would divert far too many resources away from the rest of biology.

The lure of Big Biology and the vision of the grail attracted many scientists, however, and support continued to grow and to grow quickly despite the concerns. In what might seem an odd twist, if we did not know the history, it was the Department of Energy that took the lead toward a government-funded project to sequence the entire human genome. The department had been involved with genetics studies on a large scale already, especially to assess the effects of radiation and to trace other aspects of human genetics that other researchers had largely left alone. The DOE had a leader in Charles DeLisi. Furthermore, the Los Alamos National Laboratory and Lawrence Livermore Laboratory had collaborated to establish the National Laboratory Gene Library Project, which had also developed connections with the European Molecular Biology Laboratory in Heidelberg to provide a depository for DNA sequence information. Not many facilities had the capacity to absorb the vast amounts of data beginning to accumulate, nor to process the information when it arrived. As a result, Los Alamos had proven a logical leader in data management.

A more cynical interpretation might also note that by the mid-1980s national labs were under fire for not having clearly articulated goals or a pressing need for continued funding. Some congressmen had begun to call for hearings and possible closures. What better way to ensure the continuation of the national labs than to make them indispensable, to focus their efforts on research that struck to the heart of life itself?

Undertaking a massive-scale genome sequencing project, the researchers agreed, would require three major improvements—in sequencing technology, in developing a physical map of the genome, and in data management. Whereas reports from the Santa Fe meeting had pointed to the "palpable excitement," now we read of "a palpable unease" concerning the active role of the DOE. Though doubts remained, already discussion had shifted to the best way rather than whether, to pursue a national human genome sequencing project with federal funding.

If the Department of Energy took the lead, the national labs could provide the home and support. This possibility raised questions about whether a biological project really fit properly in an environment traditionally oriented toward the physical sciences. Another possibility lay with the Howard Hughes Medical

Institute, which had already moved increasingly into the study of genetic diseases and human genetics in particular. The institute decided to hold discussions at their headquarters in July.

The National Institutes of Health or the National Science Foundation might seem the natural sponsors for such a project, perhaps jointly. Many had already argued, however, that housing the project in one place and with one agency would avoid unnecessary competition and waste of resources. Coordination and planning would be difficult enough without having to work across two different agencies and two different cultures for doing things. Furthermore, researchers worried that plopping large amounts of mission-oriented funding in one or the other of these agencies would grossly tip the balance of research there. Especially for the NSF, whose budget is relatively small and covers all the sciences, a giant genome project might well drain funding from other programs and at the least would draw attention from other basic research. This was a serious concern.

Eric Lander, at the Whitehead Institute in Cambridge, expressed reservations: "We could embark on a space-lab scale of project. But what we've been good at is devising new methods and techniques in small-scale projects. Most of the best developments of biological science techniques have been adventitious, not goal directed."[12] At a projected cost of two billion dollars or more, the sequencing project made some scientists uneasy, especially those accustomed to curiosity-driven research and to letting the peer review process determine what would be funded. A national initiative was quite a different approach to the biological sciences, they argued, and such changes are always difficult and not always good.

Discussions continued for several years while, with the enthusiastic leadership of New Mexico Senator Pete Domenici, a funding bill moved forward for the Human Genome Project in

1987. Domenici, a good friend of his state's Los Alamos laboratory, sought to locate the project and its funding there. Los Alamos knew how to handle large projects, he argued, pointing to the Manhattan Project as evidence. Unfortunately, the parallels with a "Manhattan Project" for biology have not always resonated well with the larger community; not everyone agrees that dropping bombs on civilian populations was an unassailable success, after all.

Nonetheless, Domenici pushed ahead. DeLisi decided to dedicate $5.5 million of discretionary funds already appropriated in 1987 to get the ball rolling. Meanwhile, at the NIH, director James Wyngaarden began to shed the hesitations he had expressed initially about the project. Powerful Senator Edward Kennedy took up the move to fund a human genome project, and to do it through the NIH. In the end special legislation was not required; in December 1987 Congress approved funds for the genome project for fiscal year 1988 through both the DOE and the NIH appropriations bills, with the larger amount going to the NIH.

Recommendations from an Expert NRC Committee

Before research could begin, the National Research Council (NRC) issued a report carried out with funding from the James S. McDonnell Foundation to study the subject. With Bruce Alberts, then at the University of California at San Francisco, as chair of the committee, an impressive group of biologists and ethicists carefully considered a set of questions. "Should the analysis of the human genome be left entirely to the traditionally uncoordinated, but highly successful, support systems that fund the vast majority of biomedical research? Or should a more focused and coordinated additional support system be developed

WHOSE VIEW OF LIFE?

that is limited to encouraging and facilitating the mapping and eventual sequencing of the human genome? If so, how can this be done without distorting the broader goals of biological research that are crucial for any understanding of the data generated in such a human genome project?"[13]

Note that this list did not include a question asking whether this project should be done at all. By the time the National Research Council took up the question, it was for all practical purposes decided. With the selection of a committee made up of Bruce Alberts, David Botstein, Sydney Brenner, Charles R. Cantor, Russell F. Doolittle, Leroy Hood, Victor A. McKusick, Daniel Nathans, Maynard Olson, Stuart Orkin, Leon Rosenberg, Francis H. Ruddle, Shirley Tilghman, John Tooze, and James D. Watson, all senior researchers in their fields, the decision to proceed had been made. Though Botstein, Tilghman, and Olson had all raised concerns about how such a project should be conducted, the discussion by these particular scientists clearly would focus on strategies and outcomes rather than on second thoughts. The possibilities seemed exciting.

Following an executive summary, the committee provided chapters on an introduction to the process and the science, the implications for medicine and science, descriptions of mapping, sequencing, data management, and implementation, and implications for society. This was a fine report, drawing on excellent science and providing thoughtful reflection and clear recommendations.

Pointing to advances in the science and techniques, the committee also expressed confidence that those techniques would continue to improve and that better and better tools would be available to speed the process and increase the quality of the results. The report called for having two goals, which are related but not identical: namely, that researchers should seek both a sequencing and a mapping of the whole genome. Sequencing would determine the chemical composition of DNA fragments or, put another way, which nucleotide follows which along the strand. Mapping would show where those fragments fit and how the larger genetic components are arranged along the chromosomes. Both are important, since medicine can draw on both the detailed information about particular genes or sequences of DNA and on information about whole chromosomes and the location of particular genes or sequences. This is like saying that a census should not only count all the people in an area but also keep a record of where they live. All this information is good, and even minimal resolution provides a starting point. The better the quality of the sequence data and the higher resolution the map, the more useful the information will be.

Acknowledging that the existing data banks were struggling to keep up with the sheer magnitude of information being generated, even by the mid-1980s, the committee called for a collaborative facility to collect and maintain data, as well as to oversee standards of mapping. Rather than having competing data bases around the globe, the committee said that we need a single data base. The report's conclusion displayed a prescient understanding of the significance of computational biology and the complexities of data management: "More than any other part of the human genome initiative, the handling of information and material will require organization and standardization. A single unified policy must prevail if the information is to be accurately acquired, stored, analyzed, and distributed." To establish the single facility that they envisioned, the committee called for a competitive process "in which all interested groups submit detailed applications or pilot program details."[14] This is not the way biology had usually worked, though large-scale projects in physics, space exploration, and medicine provided precedents.

The committee called for peer-reviewed projects distributed across small and large labs, therefore respecting the "long-range" approach of traditional biology. Yet its members also saw the need for coordination and collaboration, including international collaborations. Recognizing the existence of alternatives, they recommended that one major federal agency should provide leadership for the project, though they did not recommend which agency that should be. A scientific advisory board would provide guidance.

The chapter on social implications took up only five pages of the report. Nonetheless, the committee clearly recognized the potential challenges here. Commercial and legal arguments over ownership and patents, ethical concerns about use and misuse of data, social justice implications, and questions about whether and how to restrict access to information: all raised concerns that scientists and ethicists had faced before. The committee concluded its report by noting that "RFLPs [the restriction fragment length polymorphisms] will continue to be developed, maps will be made, and genetic counseling will occur even without a concerted effort to map and sequence the human genome. The greater coordination and quality control that will result from a concerted effort will in fact benefit the public by reducing the chance of misuse of poorly organized information."[15]

The committee also called for including the study of simple organisms in the initiative. Arguing that without comparison the data would be much less valuable, it recommended yeast, the bacterium *E. coli,* and the nematode worm *C. elegans*—on all of which a great deal of research had already been carried out—as excellent subjects for comparative study. As molecular biologist Shirley Tilghman, who later became president of Princeton University, reflected after more than a decade had passed, this was one of the best decisions the committee made. "Model organisms were an extraordinary investment. We learned how to sequence on these simpler organisms. And more important, we got a preview of the human genome by sequencing these organisms."[16]

The impact of the report, and the legislation and appropriations that followed, was to move the DOE and NIH to greater and clearer collaborative arrangements. Into the mid-1990s, the NIH oversaw mapping and sequencing while the DOE worked on technological innovation and data management, as well as on developing effective working relationships among all the cooperating parties.

The Project Begins

In 1988, after Congress decided to support and fund the human genome research effort, director Wyngaarden appointed geneticist James Watson to coordinate the project at NIH. At first funds were dispersed through the National Institute of General Medical Sciences, because of the presumed wide potential impact of the research. After October 1989, funds went through the project's very own administrative structure, the National Center for Human Genome Research.

Choosing Watson to head this project made sense in many ways. A passionate cheerleader for genetics research and for science, Watson was a Nobel Prize–winner and immediately recognizable by a broad range of congressional delegates and staff as well as the larger public. Highly visible and always willing to engage in public discussion, Watson had built his reputation on being a little larger than life. At first this would prove useful for carrying the genome project along.

Watson had begun his life in science as a graduate student and then headed to England, where he and Francis Crick enjoyed kicking around grand ideas and experiencing the world of science and of Cambridge. The collaboration between Watson and Crick, and the arrogant and aggressive approach taken by these two young men, led them to the double helix and their share of biological immortality. Returning to the United States, Watson found his way to CalTech and then Harvard, where he continued to aggravate colleagues and to exercise his forceful views. Many stories circulate about the divisiveness that Watson, in his enthusiasm for molecular biology and his disdain for evolution, brought to Harvard. Some are probably accurate, while others have taken on the character of urban myths. Indeed, some of those at Harvard whom Watson supposedly treated badly were quite capable themselves of arrogance and self-aggrandizement. What is important is that Watson remained a highly visible enthusiast for genomics.

In 1968 Watson was invited to direct the Cold Spring Harbor Laboratory on Long Island, New York, and his experience there strongly shaped the role he played with the genome project. The laboratory's financial situation had declined, the physical plant needed updating, and the research laboratory needed a clear mission in the face of increasing competition. With his usual passion, Watson dedicated himself to the laboratory and to securing his reputation there. The facilities today are remarkable; a scavenger hunt for all the symbolic representations of the double helix on the campus would last a very long time and yield an astonishing diversity of examples. Watson was willing (and perhaps eager) to take on the task of staring opponents in the face and telling them exactly why they are wrong. He is passionate, persuasive, and capable of real leadership. He was a logical choice to head a project that needed a clear vision to carry it forward and to continue advocating its urgency, especially in the face of all the other competitors for congressional investment.

Therefore, Watson was appointed Associate Director for Human Genome Research of the National Institutes of Health, and then Director of the National Center for Human Genome Research when that began the next year. As the Cold Spring Harbor Laboratory website puts it, no doubt with Watson's approval, "In 1992, Dr. Watson resigned his position at NCHGR after successfully launching a worldwide effort to map and sequence the human genome." Yes, the project had been successfully launched. But, as is so often the case, there is more to the story.

Part of the story is political. Watson is not a long-term Washington sort of person. He is not one to work quietly and patiently behind the scenes to build consensus and support. He is not one to give a little here to get a little there. No, with Watson, it is full speed ahead on his own terms. There are times when that is precisely and absolutely what is needed. Starting up the Human Genome Project was probably one of those times. It was necessary to have somebody arguing hard that this should be an international project while also waving the American flag to gain strong support on these shores. It was useful to have someone with such an imposing reputation, someone willing to be a cheerleader for a grand vision. Watson was the man for the job.

Yet Watson was not the man to put up with changing conditions within NIH when new leadership came along. President George H. W. Bush appointed cardiologist Dr. Bernadette Healy, the first woman director of the National Institutes of Health, in January 1991. (She remained for just two years before President Clinton replaced her with biologist Harold Varmus.) Very quickly, a war broke out with Dr. Healy on one side and Dr. Watson on the other. This surely contributed to Watson's departure from the NIH project, though he obviously was ready to return to Cold Spring Harbor, where he ruled without having to work through a seeming infinity of committees and competing stakeholders.

Owning Life?

At issue was the extremely important question, who owns the genome? And who should be allowed to profit from a publicly funded science? So many questions were involved, with so many competing interests and so many responses that might be taken as answers. On the one side lay the "high ground" of scientific freedom of inquiry. Scientists, the tradition goes, should do their research, make discoveries, and then openly share those discoveries with others. Scientific data do not really become "evidence" of anything, nor can they count as "expertise" until they are validated by additional supporting scientific results and until they are authenticated by a scientific community. Science, from this view, should be knowledge that is for the good of all and that is owned by all. Scientists should work for the public good and not for personal gain, beyond earning a living, particularly when they are publicly funded.

On the other side lay another set of values altogether. This side pointed to the potential scientific gain and argued that the development of good ideas requires investment. We need strong American companies to invest, the reasoning went, and in order for them to want to invest and develop products for the good of us all, they need to know that they can "own" the knowledge, or intellectual property, on which development rests. On this interpretation, not only is private ownership a good thing, it is necessary. Furthermore, it seemed important, in our national interest, to compete successfully with European and Japanese companies so that we could hold on to American superiority.

With the passage of the Patent and Trademark Law Amendments Act, or what was called the Bayh-Dole act of 1980, the way was cleared for private ownership of scientific information as long as it could be conceptualized as falling under the category of intellectual property or inventions. Whereas before, results of research carried out with public funding or at nonprofit universities was held properly to be the property of the people, Bayh-Dole explicitly encouraged university-industry relationships. Without such partnerships, it was argued, ideas would never be developed since companies could not afford to invest in development without being guaranteed ownership, especially in highly expensive areas like pharmaceutical development. In response, the legislation called for an increase in "technology transfer" from research scientists to the industrial and business domain.

To the dismay and even shock of many scientists, the NIH under Bernadette Healy was the first organization to seek patent protection for genome sequences produced by NIH researchers. In June 1991, the NIH sought its first patents. Scientists reacted with alarm and with excitement, and so did the public. The business world, naturally enough, experienced more excitement than alarm. Healy explained that not patenting "the mysterious gene fragments could cause American taxpayers and biotech companies to lose out on the wealth expected to flow from a cascade of genetic discoveries."[17] She reasoned that if the NIH did not patent the sequences, someone else would.

This was a clear example of what can happen in the absence of thoughtful bioscience policy and strong visionary leadership. The NIH applied for patents. The NIH received patents. As policy analyst Robert Cook-Deegan has pointed out numerous times, it does not help to point accusatory fingers at the NIH and suggest that had it not done so, the genomic data would have remained free and open to all. Indeed, the NIH's quick applications brought attention to the issue of patenting early and in the public context. By doing so it probably left open more options and created wider public awareness than if a private company had busily and quietly been seeking ownership for its own profitable interests. Why not let the public profit, some reasoned? But, of course, once the precedent was established that patenting DNA sequences is allowable, it was not only the public that profited. How could this rush to claim the goodies for Americans be justified when the NIH continued to insist that the Human Genome Project should be a fully international cooperative effort?

In a reflective article looking back on a decade of the Genome Project, John Burris, Robert Cook-Deegan, and Bruce Alberts noted that one of the biggest surprises about the project concerned patenting and funding.[18] Contributors to a collection of articles entitled *Who Owns Life?* also reflect on the developing patent policies and problems that formed and continue to shape the context for such discussions.[19] The committees that planned the project had done a pretty good job of estimating how long the work would take, how much it would cost, how quickly and how effectively new technologies could emerge, and how to manage the vast amount of data generated. What they did not foresee was patenting. Private investment had already exceeded public funding of the project by 1994. Several thousand genes and gene fragments, many thousands of full-length genes, and over one-quarter million DNA fragments had been patented.

The expert committees had assumed that such bits of data were not patentable. After all, why would the United States and then other countries decide to patent sequences of DNA or gene fragments that have no known application or functional definition? This was not an obvious outcome. In retrospect, it seems foolish that the committees consisted largely of scientists involved in research, with only a few policy analysts or ethical specialists and no intellectual property or legal experts. There is a lesson here about what should be counted as relevant expertise in policy making. Given that we cannot possibly include everyone, how can we make the best possible effort to cover the most important range of needs? And how can we ensure that we not make the same mistakes again?

From Genomics to Genomania

The ability to patent the data and to market the results—or even better, the mere promise of results—contributed to a sort of "genomania." A few dramatic developments created the illusion of great success. Nancy Wexler's team had discovered the gene that causes Huntington's disease: one gene change causes an individual to have the disease or not to have the disease. That was in 1983, before the Genome Project, but it was held out often as a kind of paradigm. First, it showed that some diseases do, in fact, originate from changes in a single gene. Second, the fact that it proved difficult to discover where the gene is located on the chromosome suggested that a full sequencing and mapping of the genome would be tremendously useful. When family studies and genetic research reveal that a particular gene is the site of a particular disease (or even if there are two genes involved, the reasoning goes), we could just look up the location on the chromosome map and find the problem.

Of course, for now and perhaps for a long time to come, that simply means that we know more. Even without the ability to "correct" a problem, that knowledge can, in itself, yield power. Knowing that we carry a particular gene that causes a particular disease or deformity allows us to make more informed choices about whether we want to have children and under what conditions. People had been making such choices for a long time, of course, but genetic tests brought more information to those choices. With late-onset diseases, we might be influenced to act differently during our lives, to take advantage of certain opportunities while younger rather than waiting until retirement. Knowing one's risk of disease has obvious practical implications.

The hope, of course, is that eventually we can move beyond diagnosis of genetic diseases to treatment. Gene therapies could take a variety of forms, from replacing the function of a particular "defective" gene to replacing "bad" genes with "good ones." Nancy Wexler, as an intelligent and thoughtful researcher who clearly cares deeply about the people she studies, fully recognizes the capacity for abuse here. Bad people can do bad things. Well-meaning people can make bad decisions. Real people are easily hurt by bad decisions. Genetic research and analysis, and perhaps genomics research and analysis of the full human genome, may lead to bad decisions.

For Wexler, however, acknowledging these possibilities does not mean that we should give up the research or that we should become paralyzed by our fears. She noted in 1992 that "If there are so many personal, social, and economic hazards and successful cures are not assured, some people ask, why proceed with the project?" Wexler answered decisively, "How can we *not* proceed?" Furthermore, she said, "I am an optimist. Even though I feel that this hiatus in which we will be able only to predict and not to prevent will be exceedingly difficult—it will stress medical, social, and economic systems that were already under a severe strain before the advent of the human genome project—I believe that the knowledge will be worth the risks."[20] She would prefer more reflective study of the genome, more informed interpretations, and greater public understanding of the science and its implications. Nonetheless, she sees the wisdom of moving forward and embraces what others came to see as the "genomania" that was provoked in part by her own successes and the leap to believe that other successes must surely be close behind.

By the mid-1990s many people were using the term *genomania*, usually derogatorily though occasionally simply to capture the excitement about genomics. Harvard biologist Ruth Hubbard, who had been in Cambridge during the debates over recombinant DNA research, emerged on the side of the critics. In this case, Hubbard pointed to the public health effects: "Not only will predictions based on 'knowing our genes' not help most people, but, this genomania may aggravate health problems by overemphasizing the health problems of individuals and deemphasizing the need for adequate social and public health practices."[21]

John Opitz, who works in developmental biology and clinical medicine, agreed. At the conclusion of a major conference on the meaning and impact of the genome project, he suggested that in the rush to genomics we were losing sight of the organism. Far from putting the "man" back in genomics, as *genomania* seems to do etymologically, it risks moving us far from the very reason we were eager to carry out the research in the first place. The "blessings" of the genome project are obvious, Opitz opined, but they are mixed blessings if they result in genetic discrimination, eugenics, or genetic fatalism within the population. Among the dangers are a reduction of attention to clinical genetics or clinical research more generally, a decrease in phenotypic and functional studies in favor of the structural studies of genetics, and a reductionist rather than holistic organicist view of development.[22]

Others have expressed concerns about ethics and distributive justice, about whether only the few will have access to genetic

WHOSE VIEW OF LIFE?

knowledge and will benefit at the expense of the many who do not. Who will have access to data about individuals or families? Who will protect rights to privacy, and who will protect the interests of those who voluntarily participate in studies only to learn later that the results will tell them things about themselves they are not sure they want to know? What do we really know about the accuracy of "DNA fingerprinting," how will we use this knowledge, and who will count as experts? There are a myriad such questions about ethical, legal, and social implications of the Human Genome Project. That is why the project was accompanied, from the beginning, by the ELSI program.

ELSI to the Rescue: Whose Ethics, for Whom?

From the early stages, the Human Genome Project included a special appropriation for ELSI research—that is, for the ethical, legal, and social implications of the project. In the beginning, as the government record shows, the ELSI working group outlined its key goals as anticipating and addressing implications for individuals and for society of the mapping and sequencing project; examining ethical, legal, and social consequences; stimulating public discussion; and developing policy options "that would assure that the information is used for the benefit of individuals and society." Over time the mission evolved. As we learned more about any one topic, new questions arose and different issues seemed more urgent.

Initially, funding for the ELSI program was suggested at 3 percent of the total genome project budget, but Watson raised that to 5 percent. Some have suggested, rather cynically, that Watson and other program leaders were just trying to buy off the critics and to lure them into support for the project. This seems unfair. In fact, Watson may well have seen the program as a way to convert critics, but not so much by giving them money. Rather, he may well have felt that exploration of the ethical, legal, and social issues would show the anticipated threats and fears to be unwarranted. Watson seems much less likely to have tried to win over his critics with financial investment and political cunning than simply to have been confident that familiarity with the genome project would calm their fears. An ingenious strategy really, and one that fits nicely the predictable pattern with most technological and scientific advances: initial apprehensions give way to excitement about the prospects and possibilities.

Watson considered opponents as ill-informed or as unjustified nay-sayers. In a paper in 1992, published just after he left the directorship of the genome project but presumably written while he was still there, he acknowledged their concerns. Yet he responded confidently, with a somewhat dismissive tone: "We need to explore the social implications of human genome research and figure out some protections for people's privacy so that these fears do not sabotage the entire project. Deep down, I think that the only thing that could stop our program is fear; if people are afraid of the information we will find, they will keep us from finding it. We have to convince our fellow citizens somehow that there will be more advantages to knowing the human genome than to not knowing it."[23]

One unexpected result from the ELSI program has been the creation of a new expertise in bioethics and a significant and lucrative bioethics industry. Before this project, *bioethics* largely referred to medical ethics with a focus on clinical medicine. Old-style bioethicists consulted, served on advisory committees, and commonly came from traditional backgrounds in philosophy, ethics, or theology. Today we have bioethics graduate and training programs to certify those who practice medical ethics in the same manner, but we also have an explosion of business for those who can engage in discussions about the ethical and so-cial implications of genetics, the genome project, and related bi-ological advances. Gradually a population of ethics "experts" is emerging who understand the issues in the biological sciences, respect the science, bring a robust appreciation for the ethical di-lemmas and conflicts in values, and have some sense of how to go about thinking analytically and practically about the range of issues confronting us.

As these individuals take on greater responsibility as our advi-sors, it behooves us to think about who they are, what expertise they have to offer, and what powers we wish to give them. Who should count as an expert on ethical and social issues, and who certifies the knowledge they generate?

Back to the Genome

After Watson left NIH, Director Healy hired Francis Collins to take up the post of director of the genome project. Collins proved to be an inspired choice. An unassuming man declaring himself a born-again Christian, who prefers bicycling to work and quiet entertainment to the high life or political intrigues of Washington, Collins holds both Ph.D. and M.D. degrees. He carries out his own active research program even while directing and defending the genome project.

In June 1998, Collins was called downtown from his labora-tory at NIH in Bethesda, Maryland, to testify before a congres-sional committee. The Subcommittee on Energy and the Envi-ronment of the House of Representatives Committee on Science asked both Collins and Craig Ventner to appear before them to discuss the Human Genome Project. In particular, they wanted to know why they should continue to invest public funding in the project when Ventner had formed a new company, Celera, with the explicitly stated goal of sequencing the genome. Why not just let the private interests prevail, the congressmen asked?

What Ventner had announced on May 9 was that he and Perkin-Elmer, the largest producer of gene-sequencing machines, had joined to form a new company that would "substantially complete the sequence" of the human genome. Furthermore, and astonishingly, they proposed to do this within three years by tackling large chunks of the genome at once rather than taking it each little piece by piece. Despite the claim that the company would produce a complete sequence, what Ventner really meant was a complete sequence of every piece of DNA that he thought was important. His team would sequence all the parts known as genes, or the estimated 5 percent of the genome demonstrated to produce proteins or to have an active role in protein synthesis. They would ignore the "junk" that makes up 95 percent of the genome, leaving that for others.

Ventner is an impatient man, but this approach made sense. The congressional committee wanted to ask, however, how Ventner's private, for-profit company would interact with NIH's publicly funded project. Did we need both, was Ventner's approach "fair," and was it good science? The NIH leadership, including Director Harold Varmus and project director Collins, both acknowledged cautiously that there was room for all. Indeed, they remained fixed on the goal of sequencing the genome, in which case productive private collaborations with the public effort should be welcome. More privately, however, they expressed concerns that Ventner would produce sloppy data, that his approach would not really work or would not have the tight quality controls that the NIH project demanded. They also expressed public concerns that the data should remain publicly available.

At the House hearing on June 17, 1998, Collins opened his

statement by calling genetics the "core science of medicine" and "the most powerful tool we have to get at the mysteries of life that still remain to be unlocked." This project to sequence the human genome is the "core of the core," the most important project in science now—even more so than the space exploration that had seemed so important earlier, he urged. Few congressmen attended the hearing, as is the way with such hearings, but those who were there obviously supported the genome project generally and were interested in Ventner's claims—they were more familiar with Collins from other hearings. They all expressed support for the sequencing project and for the NIH handling of the program to date. They were even more optimistic about the possible benefits of sequencing the genome and looked forward to the medical payoff. In addition, they seemed to accept the idea that the publicly funded science would give way to private investment in the long run.

Yet they did have questions. In particular, one congressman asked whether Ventner and Celera did not risk generating "junk science" as they rushed to get results? He declared that "science by press release" is not sound science and that we need results before declaring a victory. This clearly was a reason to continue the more cautious public project at the NIH and to support Collins. Others expressed concerns about the ownership of the data and wanted reassurances that Ventner would not rush to patent everything he sequenced. At the time, Ventner forcefully responded certainly not. His company was not greedy and would share almost all their data, except for the 100–300 sequences that they would hold as an option for patenting, in cases where there was sufficient "novelty, non-obviousness, and utility." In the context of the Bayh-Dole act's emphasis on technology transfer from public to private to promote development, the congressmen did not see this as a problem. At one point during the hearing, the two men were asked whether they saw themselves as rivals in the race for the genome sequence. Assuring the questioner that they were not, they joked that they had come wearing matching suits and ties to prove that they saw things alike.

The scientific community was not so sure, of course, and reporters asked whether Ventner's approach was more likely to be "shotgun or blunderbuss" and what he could really produce.[24] Intriguingly, instead of calling the public funding of the science into question or making the NIH group think about changing its own approach, the appearance of a race seemed to spur the researchers on and to generate even more support for the project. Thanks to Ventner's brash claims, the public was reminded of the project and of how close the goal might be. The call to action revived genomania and focused even more attention on the genes and molecular sequencing, while at the same time the organisms from which the genes were taken receded even farther from scientific focus. Even when Ventner admitted in May 2002 that much of the original research material used in his lab had come from his own DNA and bioethicists complained, the public did not seem to care. After all, the public seemed to feel, we have all that wonderful information—surely medical advances will come soon.

AAAS 2001: The Beginning or End of a Century?

When Collins and Ventner appeared together at the White House and then again under less friendly circumstances at the AAAS meeting, they were both celebrating their mutual success with the nearly completed sequencing of the genetic material in the human genome.[25] Yet they represented such different values that it is difficult to see them as collaborators, despite the similarities in their conservative ties and suits to which they had pointed in the 1998 congressional hearing. At the AAAS meeting, Ventner and Celera in particular seemed eager to upstage Collins.

Collins stood for the public funding of science, for cooperation both national and international. He promoted open publication and sharing of data, through the publicly funded and recognized standard depository for genetic information, GenBank. Collins stood for humility and scientific freedom and was at times represented as the little guy working for the cause of good and pure knowledge. Ventner stood larger and louder than life. He had the glitz, not to mention the data, some of which his group sought to patent and some of which they sought to retain for themselves while publishing the rest. Ventner was presented as the man of big business, of profit and privatism. Collins spoke of the public good to be achieved through science, Ventner urged investment in further research and development to speed the pace of discovery. Collins urged completeness, caution, and quality, while Ventner called for speed, inventiveness, and progress.

Yet this story is not really right. Ventner was the little guy, with the independent company, following the American dream, while Collins was the big government insider. Collins published his results with *Nature,* one of the most profit-oriented of scientific magazines, while Ventner published with the nonprofit *Science.* He also presented the data and its implications in exciting ways. Such researchers as New York physician and writer Gerald Weissman found the Celera system much more user-friendly than the public NIH database.[26] Instead of simple dichotomies and clashes of traditional values, the contest underscored what was new in this science. It did not fit the old, predictable patterns. Sequencing of the genome brought the end of a "race" but only the beginning of an understanding of how we might be defining life.

Ultimately, the biggest irony of all may be that this rush to sequence the human genome, with all its attendant excitement, hopes, promises, and profits, may actually lead us back to organisms. As Ventner himself has said on numerous occasions, sequencing is just the beginning. From this work in structural genomics and the compilation of vast amounts of data, we can begin to move to analysis of that data. We can begin to move to new hypotheses and new conclusions about what the data mean. Thanks to Celera's challenge, we learned that sequencing can be done much faster than ever imagined; that what count as genes actually varies a great deal—some are very large, others smaller, some widely variable and others more stable; that the genome has many copies of parts of the DNA and we do not know why; that mutation rates in the male, or Y, chromosome are higher than in the X and that the human Y chromosome seems to be the least like corresponding chromosomes in other species we have studied. It is clear that comparisons across species will prove richly valuable and that we will have to work much harder to understand how humans get to be human with only an estimated 30,000 genes, instead of the approximately 100,000 originally estimated.

A closer look at the sequences shows that there are many SNPs (or single nucleotide polymorphisms, called "snips") that serve as useful indicators of differences. If looking at a particular single nucleotide, known to be found at a particular location on the chromosome, can reveal whether a person has a genetic variation that makes a difference in function, that is extremely valuable information. And that does seem to be the case. Since we do not know much yet about functional genomics, or what the function of all those DNA sequences and genes really is, presumably future discoveries will show even more about who we are individually as we learn to interpret such things as these single "snips."

The image of DNA chips storing a record of all our snips conjures a wonderful brave new world for some geneticists, such as Lee Silver. He welcomes what he imagines as inevitable improvements, even if they present a few challenges along the way.[27] Watson also sees the developments based on our growing genetic knowledge as desirable and even necessary. He even endorses highly controversial germ-line therapy, meaning medical interventions that change the genetic material that will be passed in germ cells to future offspring, not just the DNA in somatic or body cells that will affect only the individual being treated. As Watson sees it, "It's common sense to try to develop it . . . We should be honest and say that we shouldn't just accept things that are incurable. I just think, 'What would make someone else's life better?' And if we can help without too much risk, we've got to go ahead and not worry whether we're going to offend some fundamentalist from Tulsa, Oklahoma."[28]

As *New York Times* reporter Nicholas Wade notes, germ-line and gene therapy generally offer considerable temptation, and perhaps we will hear the lure of Mephistopheles. But, he concludes, "Faust heard a good angel too; he just made the wrong choice. Having come so far and gained our first glimpse of the human genome sequence, will we really declare it too dangerous to handle? If that were our nature, we would never have risked leaving the familiar African savanna in which we rose. We can never return to its confines. We can never turn back."[29] Nor, Wade suggests, should we.

Others are much less sanguine about the short-term prospects. They are concerned about the ethical and safety issues inherent in germ-line therapies, as well as other ethical and social worries. Books and books of essays attest to the range of concerns.[30] Nonetheless, there is a momentum to the genome projects that has triumphed. Francis Collins and the National Institutes of Health—meaning human health—are busily sequencing mouse DNA. Programs are popping up in universities, biotech companies, funding agencies, and in the public and scientific imagination to pursue "functional genomics," "proteomics," and "computational biology," in particular. Soon, who knows? We may get on to "organismics."

Now, there's a thought. Let's study cells and organisms. What are organisms? What is life? What is a life? When do we know we have one? Can we define a life in terms of its genome? Can we line up all the gene sequences, look at all the SNPs and other details, and tell one person from another? Is there a "least genetic unit," and if there is does it define a life?

If so, then what is supposed to happen when we analyze a person's genome? Take yours, for example. There is undoubtedly something not perfect in your DNA that could be "fixed," as there is in anybody's. Some DNA is damaged, and some genes could be better. Let's take one of those, just one, and replace it. Any problem with that? Suppose we could do that completely safely: no danger to you, no significant cost, and considerable benefit. But, you might protest, why bother? Well, let's say that you carry a gene for Huntington's disease, or sickle-cell anemia, or any one of the others that will likely be shown to have at least some statistical correlation with a designated disease or problem. Does that make a difference?

If we say that we should not do any gene replacements under any conditions, we have a clear decision but one not likely to hold sway for long. Even if we originally think "yuk," we are likely to see the advantages in the longer run. As a society, we have accepted heart transplants, for example, and hearts seem more central to life than a single gene. If we get the replacement gene from a laboratory, cloned from another person or synthetically created, this is technology at its best. Right? What if the gene comes from another animal—a sheep, a mouse, a fruit fly? Does that matter, and if so why?

Next consider replacing two genes and taking the donated DNA from other species? What about taking all the genes from another species? After all, at least 95–98 percent of our DNA we seem to share with other primates. We apparently share many genes with all the species we know about—evolution is highly conservative. More comparative sequencing will tell us more about which genes are "ours" and which are the product of evolutionary conservation across species. If we are just talking about molecules, then the source should not matter, even if we get all of the separate genes from somewhere else. Suppose we find a copy of every gene that we need from some other organism, synthesize the ones we don't find, and then line them up in the right order (part of the function of the genome is clearly to have some genes "turn on" or activate other genes down the line). If we can get these genes all lined up, in the right order, and if they function correctly, is that a life? A human life? A good life? Does it make any sense to say that the life began with the genes—or isn't this a powerful argument that an individual life really begins only later, as development is well under way?

At this point, most people will be a little queasy. It does not seem "right" somehow. A feeling of "repugnance" starts to creep in. But why? Is it that our arrogance and feelings of species superiority make us feel that somehow we cannot simply be a product of a bundle of molecules that we have in common with, say, a pig or a rat? Even if we have a few extra, special molecules just for humans, it doesn't seem "right." But it is right if we are willing to look at life that way. If we are willing to go along with the extravagant claims and hopes of the genome initiative, we are led down this path.

What goes wrong if we accept this argument is that we are sucked into implicit assumptions about what counts as a life; we fall for a preformationist genetic view. Life is, obviously, not just a bundle of genes. We each start with our particular mix of genes. Yes, as Richard Dawkins and others have put it, the organism does serve as a way for the genome to reproduce itself. But the genome is nothing without the organisms. DNA may be interchangeable from one organism to another, or even from a laboratory dish, but DNA is just the beginning.

DNA needs an environment, full of nutrients and support, to become a life. Just as a fertilized egg needs an environment and cannot be an individual human life all by itself, so the genome cannot by itself be a life. Perhaps, as Evelyn Fox Keller put it, the twentieth century was the "century of the gene." That century is over, and gone too is the emphasis on the gene and on genomics as the path to truth and life. Perhaps, and we may hope that this is the case, we are moving into a century of the organism, an era in which we will embrace the complexities of interactions. This will call for new biology and for new policy to guide us. We have already seen moves in this direction, such as the publication of special sections of *Science* in 2002, "Whole-istic Biology" (1 March) and "It's Not Just in the Genes" (April 26).[31]

We ought not give way to an unreflective cycle of reactions, from repugnance to acceptance and excitement. Rather, we ought to seek greater understanding of the complexities and interconnectedness within a life. We must also build networks of researchers who can themselves interconnect and pursue multiple, multidisciplinary research programs. Rather than creating separate ELSI projects for genomics, we need to integrate those perspectives with all other levels of biology. We need to develop ways to embrace the human dimensions of biology at all levels, including our assessment of when a human life begins and what it means—biologically—to be a human.

Facts and Fantasies of Cloning

To those not paying attention to what was going on in reproductive science in the agricultural world—which is almost everybody—Dolly, the sheep, was a complete surprise. She burst onto the scene as if she were a character in a science fiction film. Her "creators" suddenly found themselves thrust into the spotlight of international controversy. Yet those who were paying attention should have been prepared. Indeed, the same research facility, the Roslin Institute in Scotland, had announced just months earlier that its scientists had produced lambs by transferring cells from embryos to host enucleated eggs. This was cloning, and the only thing really startlingly new about it was that it had now produced a mammal. Robert Briggs and Thomas King, then John Gurdon, and many others had shown that it was possible to transfer nuclei from embryos, even at very early stages, into host eggs to form the sorts of hybrids that are genetic clones of the donor. That part was not new.

Crossing what was commonly regarded as a mammalian barrier really was new, and for many it was shocking and frightening. When the story about cloning frogs had appeared, the news media reacted with some excitement at first, then lost interest. Since even the earlier cloning results of 1997 were about sheep and drew on techniques and approaches that others had been pursuing for decades, and since the Roslin Institute had not made a fuss about the results, that frenzy died quickly. With Dolly, however, things were different.

Dolly was born in late 1996, a healthy lamb, but the researchers took their time writing the paper announcing her birth. They wanted to present the results in approved scientific format so that others could follow their procedures and reproduce the experiment. Their sponsor, PPL Therapeutics, Ltd., wanted to seek a patent on the cloning techniques used, and that took time. They eventually submitted the paper to the British journal *Nature,* one of the two most prestigious scientific journals, and after an expedited review process the paper was accepted January 10. *Nature* recognized this as an important story and moved the paper into print for the February 27, 1997, issue.

Nature, like other leading scientific publications, puts out a press release on the Friday before publication date. The stories are officially "embargoed" until a few days before the actual publication date, and the extra lead time is intended to allow reporters to study the issues and to write better-informed stories when the embargo is lifted. Obviously, the science journals also hope to generate more publicity for their stories by giving this heads-up. Usually, this approach works well, and since science writers are typically expected to cover all fields of science and cannot possibly be experts on every new story that comes along, they appreciate even a little bit of lead time. Occasionally a story strikes reporters as important, or the release reaches the desk of an eager editor or an unscrupulous "leak," and the embargo is breached.

This happened with the Dolly story. On Thursday, February

20, a message went out by email from *Nature* to science reporters. The brief explained that "The lamb on this week's cover was raised from a single oocyte (egg cell), whose nucleus had been replaced with that from an adult sheep mammary gland cell. It may be the first mammal to have been raised from a cell derived from adult tissue." The release continued: "The implications of this work are far-reaching."[1] On Friday, reporters had access to the paper itself.

Science and the Press

On Sunday, the British *Observer* published a story by their lead science writer, Robin McKie. Even the Roslin Institute later acknowledged that McKie's story was as balanced as possible and "resisted taking a sensational line." McKie later claimed that he had relied on his own sources and did not technically break the *Nature* embargo. The Roslin researchers realized immediately that this story was big, and they set to work to meet the media frenzy. They noted that Dolly Parton told them she was "'honoured' that we had named our progeny after her and that there is no such thing as 'baaaaaed publicity.'"[2]

While the researchers applauded the *Observer's* restraint, they did note that McKie's story raised the specter of human cloning and expressed concern. Dolly's American debut, in the late edition of the Sunday *New York Times,* was less restrained. Gina Kolata broke the story after reading the *Nature* press release and talking to her editor. They decided that they would have a story ready to run in case anybody else broke the embargo that stood until the following Wednesday, but they would not be the ones to break it themselves. Given that this was just-breaking news, Kolata had to write quickly and to be ready for anything. When nothing had happened by late Saturday, she assumed that the

WHOSE VIEW OF LIFE?

embargo had held, giving them at least until early in the next week to follow up. Nonetheless, she prepared a story from what material she had in hand that could run as needed, and suddenly it was needed for the Sunday issue when the *Observer* broke the embargo and published its account.

That front-page Sunday story is highly problematic. Kolata is a serious and typically thoughtful science writer, then regarded as one of the best. Yet, perhaps because the story was so rushed, or perhaps because the *Nature* paper itself was so restrained and dry, Kolata adopted a popular and speculative approach. Arguably, the way she presented the story—Americans' first exposure to the idea of cloning—affected what happened after.

Kolata presented very little science, perhaps because the relatively short and terse *Nature* paper by Ian Wilmut and his colleagues did not provide much detail or perhaps because she wanted to capture the larger picture and the implications of this research. She did, however, find a few "experts" to interview even on short notice over the weekend. Naturally enough, she contacted individuals who had previously commented on genetic and molecular biology and its implications, as she wanted to find strong spokesmen to provide powerful quotations on extreme positions. Princeton biologist Lee Silver was one of those. Kolata noted that Silver had been finishing his book *Remaking Eden* and had been in the process of explaining why such cloning is impossible when he got the news. Just in time, he noted, to make the necessary revisions. With apparent glee, Silver commented: "It's unbelievable. It basically means that there are no limits. It means all of science fiction is true. They said it could never be done and now here it is, done before the year 2000."[3] Those of us who have spoken with reporters, only to have what we say misunderstood or distorted, might imagine that Silver really could not have said anything quite so extreme or silly. But, no, he has repeated the sentiment in print more than once. The choice to quote Silver, and in just this way, clearly had the effect—whether intentional or not—of inflaming passions.

Despite Kolata's quotations of sensible and balanced thinkers like Lori Andrews and the scientist/creator Ian Wilmut himself, the images that stand out are the extreme ones. Thus, readers of the Sunday story were hardly likely to remember Wilmut's measured calls for caution and recognition of the potential for misuse of this new technology. Instead, they would probably remember the words of medical ethicist Lewis Munson: "The genie is out of the bottle. This technology is not, in principle, policeable." Furthermore, Kolata recounts an idea that Munson once had for a story "in which a scientist obtains a spot of blood from the cross on which Jesus was crucified. He then uses it to clone a man who is Jesus Christ—or perhaps cannot be." This image just sits there, undigested and uninterpreted. Papers and reporters across the country picked up the image, without appreciating the subtleties that Kolata herself undoubtedly intended. Is that fictional man Jesus Christ? Obviously not, and we should realize that cloning cannot work that way, but Kolata did not point this out.

The next day, Kolata gave us another story, this one focused on the ethical issues explicitly. Silver and Munson are again quoted, but so are those who consider what cloning really means and who point out that "replication is not procreation." We are not, however, told what that means. It is reasonable to assume that Kolata, like other reporters, was doing her best to report a hot story, but she presented little in the way of information or real insight into just precisely what biology was involved. Nor did her story or others provide a deeper discussion of what ethical and social implications might be at stake or how to interpret them. An editorial the next day acknowledged the uneasy feeling that we all had on hearing about Dolly, a sheep cloned from a technique designed to improve agricultural output but with far-reaching and very uncertain implications for human applications.

The main story in the *Times* came a few days later on March 3, written as a special report by Michael Specter and Gina Kolata. This long and excellent presentation explained in much greater detail the science involved and addressed the implications of the research. This was fine reporting and well worth re-reading and saving as a model of what scientific journalism can be. Rick Weiss's reports in the *Washington Post* set a similarly high standard.

In the context of explaining the scientific contributions and the technical advances, Specter and Kolata also presented the lead scientist, Ian Wilmut, as a thoughtful, decent, and caring man, wearied from the ill-informed media frenzy. Wilmut said that he was too tired of hearing "the worries of everyone from the Vatican to President Clinton to muster any further outrage." He felt sad that people were rushing to the conclusion that they could use cloning to get back their dead children. "It's heart-wrenching. You could never get that child back. It would be something different. You need to understand the biology. People are not genes. They are so much more than that."[4] Such thoughtful reflection should have made a difference in setting the story right and helping those with good will to understand what was and was not scientifically involved; in turn that understanding should have helped to stem the tide of concerns about replicating humans. Of course, some newspapers and magazines sensationalize every story and thrive on excess. Even for those serious reporters who try for a more balanced view, however, the damage had already been done. In the face of widespread biological illiteracy, few writers had sufficient expertise to understand the science without interpretation, and in the face of explosive public reaction, images of unending replicates of this nice but shocking sheep—and us—were apparently just too tempting.

This story raises questions about the role of the press in presenting and interpreting science. In the United States today, the National Association of Science Writers does a good job of providing information and a network of writers to help keep each other informed about what sources are available and what the hot issues are. In the 1920s, the Science Service was a leading source of information about science stories. Edwin E. Slosson, a chemist and also a journalist and editor, was selected as the first editor in 1921. This publication arose when philanthropist E. W. Scripps developed a collaboration with the American Association for the Advancement of Science, the National Academy of Sciences, and the National Research Council to provide a source of reliable scientific information and to popularize science for a wide lay audience. The historical records of this important service, now available through the Smithsonian Institution Archives and its website, show that it used science to make social arguments. For example, it came out in favor of evolution during the Scopes trial. The record, viewed historically, shows much more clearly than is evident today, when we are in the midst of dramatic events, just how much media presentation matters to public reception of ideas and their applications and implications.

Of course, I am not claiming that Gina Kolata and the *New York Times* alone caused the American uproar about cloning, but neither did the early reporting help dampen the overreaction. *Time, Newsweek, U.S. News and World Report, The New Yorker,* and newspapers and television stations around the country called up images of brave new worlds and duplicate animals and people. Canon advertised its photocopy machines with pictures of two sheep looking just alike with the slogan, "We've been making copies for years." The researchers at the Roslin Institute were overwhelmed with email, phone calls, and hoards of visitors urgently insisting on access. Visitors wanted to see and get pictures of Dolly, to see and get pictures of Ian Wilmut, to hear more, and to write more and more dramatic stories.

Most interpreters recoiled emotionally from imagined "photocopied babies." Meanwhile, serious scholars and careful journalists struggled to explain that replication at will is not at all what is really at issue. Many factors influenced the way the story was received. To appreciate the significance of the responses, however, we first need to look more closely at the science involved.

The Science

Recall that nuclear transfer, which is what Briggs and King had done with embryonic frog cells, involves taking a nucleus from one (donor) cell and transferring it to another enucleated (host) cell. But it is not quite so simple. First, the host cell has to be prepared. This has to be an egg cell, which would normally begin dividing when it is fertilized and which has the apparatus to divide and to grow. It took quite a lot of trial and error to discover that an egg cell works best to support the transfer if it is still in the earlier, or oocyte, stages and not a fully mature egg. When the researchers remove the nucleus from this egg cell, in effect they hollow out its nuclear and therefore most of its genetic material. Though some genes are nonetheless left behind, such as mitochondrial DNA and RNAs from the mother, the bulk of the nuclear DNA is removed.

For simple nuclear transfer, the researchers would then take a nucleus from a donor cell and transfer that nucleus to the host egg. But removing the nucleus from the donor without taking along some other material and without destroying the nucleus itself proves difficult, especially in higher organisms like mammals. Therefore, Wilmut and his team concluded that they should try transferring the entire cell, including the nucleus. By this method they would take all the inherited nuclear material and some extra. Perhaps the extra material in the cytoplasm of the donor would confuse things, but the technique was so much simpler and cleaner than actually removing the nucleus that they decided to try.

It worked. In 1995, Ian Wilmut and his colleagues decided to take the donor cells when they were in a normally quiet stage, not undergoing cell division or other developmental activity. They transferred these cells, including the nuclei, into the hollowed-out host oocytes, which produced a sort of chimeric hybrid. Then, to get things started again, they administered a small electrical shock, which has the effect of jump-starting the cell division process. In their 1995 experiments, they transferred 244 nuclei, which developed into thirty-four embryos that were each transferred to the uterus of a surrogate mother. Of these, five lambs were born and the two named Megan and Morag survived to healthy adulthood as the first mammals ever cloned from differentiated cells. These two came from differentiated but not adult cells.[5]

Remember that Ian Wilmut and his team worked at the Roslin Institute in Scotland, and they were paid to do agricultural research in order to improve productivity. Genetic engineering had proven highly valuable for "pharming" and promised to offer even more great advances. Just imagine being able to splice in genes that would protect the animals from disease, for example, or others that would enhance milk production. Splicing genes

　　　　　　　　WHOSE VIEW OF LIFE?

into single eggs is difficult, but splicing them into other cells and then transferring those cells to oocytes held great promise. The researchers did not focus on humans or even have humans in mind. Wilmut made very clear from the start that he, at least, sees cloning humans as immoral. For Wilmut, his work really was very much about sheep. This attitude probably allowed the researchers to continue probing for better techniques even after they had achieved what others might have regarded as the earth-shattering news of Megan and Morag. Most people never noticed these two.

Why not continue by trying nuclear transfer from other cells, like adult somatic (body) cells? They then used nuclei from three kinds of cells: some from embryonic cells; others from fetal neuroblasts (cells that normally give rise to nervous tissue), since those cells normally are relatively easy to culture; and some from adult mammary gland cells. Hundreds of experiments resulted in four live lambs from the embryonic cells, two from the neuroblasts, and one from mammary glands. That is, only one (Dolly) came from the adult, differentiated cell. The results paralleled earlier findings in showing that the later the stage the donor cells come from, the more difficult the cloning.

For scientists, the astonishing thing was that it worked at all. Never before had a differentiated adult somatic cell (that is, one from the body of the animal and not from the germ line or an embryo or fetal tissue) been cloned. Furthermore, as Lee Silver suggested, most scientists had long since decided that it could not work. Like Briggs and King, they had concluded that adult somatic cells had undergone too much differentiation to "turn back." Somehow, with their particular technique, Wilmut's group had gotten the cell, in effect, to dedifferentiate and become reprogrammed—at least, that is how they interpreted the results.

Cloned in 1996, Dolly did not appear on the public scene until the *Nature* paper was in print and ready for consumption in early 1997. By later that year, the researchers had taken another step: using fetal fibroblast cells (cells from the fetus that give rise to other kinds of cells as development progresses) that had been genetically engineered to carry an additional gene, they produced another clone, Polly. The gene carried the ability to code for a human blood-clotting factor, a potentially valuable pharmaceutical protein. They published these findings in rough form but, given the pressures to patent and to develop the commercial side of the work, they have kept secret the more recent studies funded at the Roslin Institute by a sequence of partners and owners, as is true elsewhere. After this initial flush of amazement, or the "bolt from the future" as several reports called the announcement of Dolly, the focus outside Scotland turned quickly from sheep to humans.

Discussions quickly escaped the scientists' agendas, of course. Wilmut has made heroic efforts to help keep things in perspective, to insist that Dolly is just a sheep, that he considers cloning humans for reproductive purposes quite unethical and unsafe, that this is a technology that needs considerably more work and refinement before we would even think of using it for humans in any case. As the designated point man for the press, he willingly if not always eagerly stood on the front line, answering questions and patiently explaining for the gazillionth time some basic fact of the matter that yet another questioner had confused. What shaped the media and public frenzy of Dollymania? Surely, along with the general background of scientific illiteracy that leads the public to rush to misguided misunderstandings, and along with the rarity of solid media presentations, important features of the climate surrounding science shaped the reactions. In particular, genetic determinism, debates about what counts as "sound science," the rise of the bioethics community, and political responses all played important roles.

Genetic Determinism Again

When the Human Genome Project was the scientific news story of the day, so many people had become so focused on the genes and on the genomes that they had forgotten about organisms. They had forgotten about the environment in which those genes matter, and they definitely had forgotten about the intricacies of development and differentiation. So many who enthusiastically entered the genomics race were trained as molecular biologists, steeped in molecular techniques and the rituals of the laboratory. It hardly mattered to them whether they were studying fruit flies, nematode worms, sheep, or humans. Some of these specialists had studied little general biology, and they knew little about organisms, nothing much about evolution, and nothing at all about ethics and the social impact of science. They were experts, and their expertise was molecular genetics.

There is nothing intrinsically wrong with that, except that we also need experts who can translate across different fields of expertise. And we need somebody to remind us that genetics is only a small part of life. As Wilmut and other leading biologists, such as Richard Lewontin, noted over and over again, patiently and then perhaps somewhat sadly as people seemed so unwilling to listen: life is not genes; life does not begin or end with genes; a genome does not define a life; nobody can replicate a life. Or as philosopher of science Philip Kitcher reminded his readers in various contexts, "there will never be another you." It matters very much to get this right.

Immediately, nonetheless, the public and the press jumped to conclusions. Clones were copies, genetic copies, so therefore they must be nothing but copies of other organisms. That meant, they rushed to conclude, that a clone was a genetic twin. It is, however, a time-delayed twin, since one of the twins would be more advanced in development than the other, perhaps even old enough to be the parent of the other. The child is not supposed to be the genetic brother of the man. Conceptualized that way—that inaccurate and wholly misleading way—cloning seems most definitely odd. Indeed, it seemed abhorrent to many, who argued that a cloned individual would lose his or her autonomy. With the current socio-political focus on genetic determinism, even some scientists who know better talk as if genes do define us, as if two persons having the same genome must in some sense be the same individual. This notion appears to threaten autonomy and makes people nervous. Responses of this sort were not just the willfully ignorant reaction of an uninformed public, because we can point to scientists who said similar-sounding things about the power of the genome during arguments for public funding of the genome project.

Molecular geneticists and biochemists who knew better scientifically lobbied hard for the central significance of the genome, as if knowing gene sequences would solve all medical problems. Other molecular geneticists who may or may not have known better joined in doing the research. Some probably really do believe that genes are basic in defining each individual life, as Lee Silver seems to and as Richard Dawkins suggests in most of his genes-and-evolution-are-us arguments. It is not surprising, therefore, that the public should jump to conclusions and equate the copying of genetic information with the copying of people.

Some scientists who should have known better and who could

have helped instead responded to cloning by emphasizing that twins are more alike genetically than clones, since the clones differ by the mother's mitochondrial DNA, after all. But that argument missed the point and retained the misguided emphasis on the genes themselves. Life is not about genes alone. We cannot say that often enough. Thomas Murray, who as president of the Hastings Center for bioethics in New York was invited to testify in a congressional hearing, understood that. Though distraught that his own daughter had been murdered only months before, he recognized that cloning could not bring her back. In an editorial, he spoke with great wisdom about the biology of humans. "About the only thing we can be certain of is that we are much more than the sum of our genes." For "Each of us is a complex amalgam of luck, experience and heredity. Where in the womb an embryo burrows, what its mother eats or drinks, what stresses she endures, her age—all these factors shape the developing fetus. The genes themselves conduct an intricately choreographed dance, turning on and off, instructing other genes to do the same in response to their interior rhythms and to the pulses of the world outside. How we become who we are remains a mystery."[6]

As Murray said in his March 5 testimony to the House Subcommittee on Technology, "good ethics begins with good facts." Strikingly, even some of the molecular biologists who were such enthusiasts for the genome project were not well versed on the developmental facts. In addition, many prefer not to discuss such issues as the social or ethical implications of cloning or stem cell research, saying that they are scientists who work in the lab and produce "pure" and "objective" knowledge. It is, these scientists maintain, up to others to interpret how to use the science. That seems fine, but if every scientist is like this all the time in all situations, we are doomed to have decisions made by those who are scientifically ill informed or even scientifically illiterate.

Sound Science: Who Says?

Most scientists have resisted taking the lead in debates over definitions or implications not deemed centrally important to the science. As a result, they have allowed the public to define the central terms and framework of discussion. As a television producer reportedly told Lee Silver at one point during the public debates about cloning: "Clones are not what you think they are." At first Silver objected, but then he concluded that "Science and scientists would be better served by choosing other words to explain advances in developmental biotechnology to the public."[7] Yes, scientists should obviously choose terms that make scientific sense, but they should not avoid using the "real" language that scientists themselves use. Rather, they should work to provide clear definitions and be prepared to inform public debate. They should do this individually, and the scientific community needs to work collectively, as the Science Service and its supporters did in the 1920s and 1930s, to continue informing the debates.

Scientists should neither retreat into their labs and leave interpretation entirely to the public, which is then doomed to making decisions without information, nor arrogate unto themselves alone the right to make all such decisions. Rather, we need scientists and the scientific community to help interpret the science and its implications for the public. Scientists should sit at the table along with others to negotiate decisions about how, when, under what conditions, and to what ends to support scientific research and its applications. Without those informed scientists' perspectives, we risk making bad policies, in some cases without even realizing that we are doing so because we thought we were talking about something else.

Yet neither should they go too far. Richard Lewontin acknowledges that this is a delicate balance. "Far too little attention has been paid to the immense damage done by the propagation of false ideologists and false metaphors by scientists. What scientists say about the world is at least as important from a moral and political point of view as the actual state of nature."[8] Robert A. Weinberg, a 1997 National Medal of Science winner, agrees. He argues that scientists have no more special claim to deciding moral issues by themselves than the informed man or woman in the street or than "a senator from Kansas or a cardinal in Cologne."[9] Furthermore, scientists need to be careful not to oversimplify because they assume that the public cannot understand. An Australian developmental biologist got himself into trouble when he lumped embryonic stem cells and embryonic germ cells together in explaining the science to parliament. He noted afterwards, with some regrets: "Embryonic germ cells had never been explained to these parliamentarians before, so I simplified and just called them embryonic stem cells." He acknowledged that "This is not absolutely correct, but they are embryonic and they are stem cells and you can't tell the difference between them." Nonetheless, he admitted that he made a mistake and urged other scientists to consult with professional organizations if they were going to testify or talk to the wider public.[10]

Notwithstanding the many efforts to explain that genetic replication does not mean that the resulting offspring cells, tissues, or organisms will be exact copies of the original, the popular perception still sees it that way. As Silver noted, cloning escaped the biologists because of Dolly and the subsequent furor and took on its own public meaning. This does not mean that biologists should give cloning up to the popular press without trying to communicate their own knowledge and experience.

The call for scientists' engagement in public debate about science, as experts but not as the ultimate decision makers, may or may not fit with assumptions on Capitol Hill or in the regulatory arm of government. It is intriguing to note that the conservative movement to develop a "Contract with America" in Congress, led by Congressman Newt Gingrich, featured an enthusiasm for science along with a distrust of experts. This led to an odd tension in Congress that played out in various ways. With Gingrich's rise to power, he held increasing sway with the Republican majority leadership. When Gingrich ridiculed some idea that he did not like, he called for what he saw as "sound science" to settle the question. Starting from a few cases, by the 105th Congress fully twenty-one introduced bills had the exact phrase "Sound Science" in the title; by the 106th Congress the number had risen to twenty-eight, and at least thirty-eight in the 107th. What does this mean?

Often what "sound science" really means to politicians is "science that supports my particular interpretation and leads to the conclusions I want." At times the explicit emphasis on science that is sound has led to demands that scientific experts offer merely "facts" and no interpretations or "fictions" based on but going beyond the scientific data. This point keeps coming up in congressional hearings and discussions of global warming, for example, and it comes up with debates about cloning and stem cell research. What is the science, members of Congress ask? When scientists disagree, what is the "sound science," and how do we identify it? What do we do when science changes over time, as it inevitably will? Who are the experts who will certify the knowledge?

As president of the National Academy of Sciences, Bruce Alberts has been called to testify on numerous occasions. In discussing evolution and science education at one point, he was very clear and effective in explaining that good science and sound science depend on the scientific claims' having been confirmed and accepted by the larger scientific community. Yes, out of 10,000 scientists you may find a few who will say "not x" when all the others say "x," and it is not then rational nor good science to say that we must find a balance between the two positions. In such a case, "not x" is not a reasonable position—not now, not on the existing scientific evidence, though that could change in the future and may well have changed in the past. Therefore, evolution is good science and should be taught in the schools, Alberts concluded definitively.

Or he might have said, we should not assume any sort of ultimate genetic determinism but should recognize that organisms develop through a complex process that only begins with genes. This means that cloning will not be copying, either of people or of sheep. But this scientific fact does not say anything about how we should use the science, or whether we should actually proceed to clone humans. That is another matter altogether, and one that calls for expertise in addition to that of scientists. As even arch-evolutionist and genetic enthusiast Richard Dawkins says, "Science and logic cannot tell us what is right and what is wrong."[11]

But then, who are the experts who can advise us on complex issues that strike at the interface of bioscience and society? Are bioethicists the proper experts? Are they the ones who should tell us the "pure" and "objective" truth about what is good and bad, about what we should do?

Bioethics on the Rise

When Louise Brown was born through *in vitro* fertilization, the media were hungry for news, but the deed had already been done. There was the healthy, normal, and happy baby with normal and happy parents, proving that her "creation" was not evil. How could there be any serious challenges to the ethics of bringing such a child into the world, one might well have asked? Besides, we were bioethical innocents in those days. There was no corps of trained bioethics experts ready to pounce on every issue of biology and medicine and analyze the ethical implications inside out. By 1997, however, there was. There was room for debate about the ethical appropriateness of cloning humans. There wasn't any happy baby yet. The placid face of a sheep, no matter how apparently contented and appealing, just did not evoke the same gut feeling that "this is a good thing for us humans."

The cadre of bioethicists came into being in part because of the ELSI program within the Human Genome Project, which provided money to researchers who would engage issues of bioethics and train others in bioethics related to genetics and genomics issues. Yet, as always, there is more to the rise of bioethics than the opportunities presented by ELSI funding. In 1947, the Nuremberg trials had yielded a code to guide medical research with human subjects and prohibit Nazi-like human "experimentation" that quite clearly violated human dignity and individual rights. Different medical groups in numerous countries developed their own ethical codes as well. Among the principles designed to protect individuals from abuse was the requirement that "The degree of risk to be taken should never exceed that determined by the humanitarian importance of the problem to be solved by the experiment."[12]

In the 1960s and 1970s, the NIH continued discussions on protecting human participants in research and medical trials. This line of inquiry held more than just abstract policy implications but also practical concern for the NIH itself, as its researchers began carrying out more and more research on their home campus in Bethesda, Maryland. In 1974, the National Research Act became law, thereby establishing the National Commission for the Protection of Human Subjects of Biomedical and Behavioral Research. The commission's charge included identifying the "basic ethical principles that should underlie the conduct of biomedical and behavioral research involving human subjects" and also developing "guidelines which should be followed to assure that such research is conducted in accordance with those principles." What became known as the Belmont Report followed, coming out of a conference held at the Smithsonian Institution's Belmont Conference Center. This report provided a set of guidelines and also a framework for further discussion by pointing to basic ethical principles. Any actions must start from a solid foundation: "Three basic principles, among those generally accepted in our cultural tradition, are particularly relevant to the ethics of research involving human subjects: the principles of respect of persons, beneficence and justice."[13]

June 18, 1991, brought an unusual collaboration in Washington, as a significant majority of federal agencies jointly published an agreed-upon "Common Rule" to regulate the conduct and support of research throughout the government. This rule provided the framework, for example, for establishing internal institutional review boards (IRBs). Various iterations to interpret and expand the rules have since clarified some areas, while others lying outside the clear jurisdiction of the agencies have remained open to interpretation. Cloning falls into the latter category.

Assessing what counts as excessive risk or the best way to insure preservation of the basic ethical principles is precisely what is at stake in the most contested bioethical issues. What we should do with our increasing knowledge of genomics and whether and when we should attempt gene replacement therapies or genetic enhancements or germ line therapies: these are all challenging questions. When might cloning prove to have sufficient benefits to outweigh the risks, and when do the risks prevail? In the United States, we have begun to rely on bioethicists to help us decide.

Perhaps the most publicly familiar bioethics guru is Arthur Caplan, director of the University of Pennsylvania Bioethics Center. Critics ridicule Caplan's apparent willingness to express views on such a wide range of subjects, and they laugh at seeing him on multiple television stations at the same time. "Real scholars" should not be so public, they scoff. Yet fair-minded citizens must admit that Caplan has been willing to take on tough ethical questions when others were not. He has helped to develop graduate training programs, to train ethicists to talk to the press and to the public, and has urged that bioethicists should study the science and the medical impacts that they are talking about before they start talking. He may be on every news station once in a while, when a hot issue pops up, and his website at the University of Pennsylvania came to receive 17,000 hits a day rather than the 500 a month it received before Dolly.[14] That is not necessarily a bad thing. If the public, reporters, scientists, religious leaders, teachers, and other scholars are eager to have answers for themselves or their audience, constituents, parishioners, or students, then it is good that someone is making the effort to satisfy that hunger for understanding with as much in-

WHOSE VIEW OF LIFE?

tegrity and as much expertise as possible. We all wanted to know what the science was about, yes, but also what it means. Caplan and other leading bioethicists at centers around the country perform a service in stepping up to help us gain perspective.

More problematic are the self-proclaimed bioethicists and theologians-turned-bioethicists who add their voices to the mix in reaction to popular presentations of "problems" generated by scientists. Many of these advocates are not seriously willing to engage in reflection about the difficult challenges of sorting out right and wrong. Rather, they begin with *a priori* convictions about what the conclusions should be. They hold particular views that they want to promote, whether they are knee-jerk reactions, deeply held absolutist beliefs, or misunderstandings. Some reveal a willful disinterest in learning more about the science or the medical experimentation. These self-proclaimed ethicists insist that they know what is true and right and good, and they will dedicate themselves to its advocacy no matter what. This is disturbing. This is, however, where the chairman of the President's Council on Bioethics, Leon Kass's, "wisdom of repugnance" may lead us, for he relies heavily on his own intuitions, on his assumptions that our intuitions will match his and that if they do not match there is something wrong with us.[15]

Intuitionists often get into trouble at just this point. They rely on intersubjectively shared intuitions as their source of certainty and a sort of truth. Yet history shows that often our presumed certainty about repugnance is replaced later with acceptance. Often our intuitions are just based on prejudices, which grow out of our limited experiences and our particular contextual values. Intuitions can provide a useful starting point for moral reflections, but they are a bad foundation for social policy when they lead to intransigence, absolutism, and resistance to revision in the face of new evidence. Surely, we ought to learn from history that relying on intuition as a guide to ethical behavior is un-wise. Nature evolves, science evolves, social attitudes evolve, and our responses should be expected to evolve.

This is not the place to reiterate all that has been written on the ethical issues raised by cloning. Many books wait at your lo-cal bookstore to tell you more than you ever wanted to know about virtually every available bioethicist's opinions on the sub-ject. Indeed, many edited volumes include the same essays by the same people, because those points of view are thought to be so useful and so marketable that they bear repeating to other audi-ences. This display of excitement about the ethics of cloning re-flects the rising numbers of bioethicists as well as their increas-ingly public role in society.

The most important questions have fallen into several clusters. First came questions addressing issues of autonomy, person-hood, individuality, and human dignity. If a clone is a genetic copy of the donor cell, ethicists cried, it will not be an individual; it will not have autonomy and as such will be denied human dig-nity. Yet, as noted earlier, Lewontin and others have publicly pointed out clearly that this is nonsense since "replication is not procreation." The clone is not even a replica of anything except that the nuclear genes are literally the same as the donor's. Add those to the additional genes of the host egg cell, and mix with all the developmental processes and environmental variations, and the offspring will be far different.

Indeed, as commentators delighted in pointing out, a clone would be less genetically like its "parent" than twins are like each other. In a particularly compelling congressional hearing early in 1997, NIH Director Harold Varmus noted that he was one of twin boys. He and his brother made the point that no two individuals, even identical twins, are really alike. After all, he told us, he was the head of a major research institution in the

United States, while his brother was head of a major research institution in Canada. Very different, he joked. Then, having the rapt attention of his overpacked audience, he tried carefully to make the point that clones are not copies and should not be seen as such.

Nonetheless, for those committed to the interpretation that an individual life begins at conception—an event that for many includes an act of ensoulment—cloning causes problems. When does the clone get his or her soul? When does a life begin: is the placement of the donor nucleus or cell into the host equivalent to the moment of conception, and therefore the beginning of life? In fertility clinics, if an egg is fertilized by "normal" means and then caused to divide, is the procedure a violation of life because by splitting the one into two it divided a soul? How is this different from what happens in natural twinning? What if the technician intervenes to create twins out of what would have been one and then freezes one of the "twins" for a later day, thereby creating time-delayed twins—as cloning does? Is that different, and is it substantively worse than normal twinning? What, precisely, is at issue here?

These questions are the wrong ones for those who do not imagine that an individual person's life begins in any meaningful sense at fertilization. For those with the conviction that a life does, inevitably and unquestionably, begin at the coming together of an egg and a sperm, these are legitimate and important questions. They must ask, for example, as bioethicists Glenn McGee and Art Caplan have, "what's in the dish" at the fertility clinics if it isn't a life because it was created in a different way?[16] For those who begin with such convictions, they must not pretend that they are appealing to any biological facts to support their claims. Rather, their convictions must have other grounds, some outside source that reflects matters of convention for people they are trying to reach. For all of us, these are fair and profound questions. Cloning does strike at the heart of much that we hold dear.

Second comes a cluster of concerns about the scientist's hubris in doing such experiments at all. Were the scientists who created Dolly "playing God" in a thoroughly unacceptable way because they interfered with nature? Were they transgressing boundaries of nature that would get us into trouble? Reproductive scientist Roger Gosden takes up this concern in the larger context of reproductive changes and challenges generally, laying out a range of public hopes and fears as well as the medical prospects.[17] Does anybody really care about sheep, or do we feel that it would be unacceptable to do for humans what Wilmut and his team had done for sheep? Some of these questions became practical concerns as well as philosophical ones, since it would be unethical to do things with humans that we are not sure are safe. Molecular biologist Rudolf Jaenisch and Ian Wilmut made this argument when they wrote that "attempts to clone human beings at a time when the scientific issues of nuclear cloning have not been clarified are dangerous and irresponsible."[18]

Furthermore, does cloning involve transgressing some natural evolutionary boundaries? And do such transgressions bring considerable risk of unexpected disease, deformities, or degradation of the gene pool, for example? Whereas ethical discussions relating to the Human Genome Project had concentrated on questions about whether we were mixing genes in unacceptable and excessively risky ways, here the question was whether it might not be unethical and risky to mix genes at all. Unlike normal, sexual reproduction of humans, which brings together the genetic contributions from two different parents and lets the combination work itself out in the developmental process, cloning transports the genetic contribution from one parent almost directly and intact to the offspring. Especially if the donor were of a significantly different age than the host egg cell, as would happen with somatic cell nuclear transfer from an adult, cloning then seemed "unnatural" and perhaps risky if it led to abnormalities and deficiencies.[19]

Others raised important issues of social justice, many of which had been raised about genomics. If we invest so much public funding in a scientific project, how will it serve the public? What else should we be doing with that money instead? This is one argument against funding such research, but not against the research in itself. Others noted that once we have developed technologies and medical procedures that are very expensive and must necessarily be limited to only a few, there will be unequal access to those "goods." This raises issues of distributive justice, which hold for any technical advance and are not peculiar to cloning.

Among the many voices, Philip Kitcher and Ronald Green provide quite different and particularly thoughtful perspectives on issues of social justice and rights in a democratic society confronted with technological and scientific innovations. Green points to what he sees as the legitimate right of couples to have children, and he urges that we at least not close out their options because of vague and ultimately unsustainable moral objections.[20] He calls for more careful consideration of their interests and that we not allow the absolutist demands of pro-life groups to foreclose this option. Kitcher offers strong arguments that our confused response to cloning reflects a larger failure to develop any clear directive policies or moral directives. Especially with scientific advances, which are foreign to most policy makers, developing such thoughtful and intelligent guiding policies is a challenge. As Kitcher puts it, we have a moral obligation to improve people's lives and to develop wise social policy to do so.

"What is truly shameful," he concludes, "is not that the response to possibilities of cloning came so late, nor that it has been confused; it is the common reluctance of all the affluent nations to think through the implications of time-honored moral principles and to design a coherent use of the new genetic information and technology for human well-being."[21]

Constitutional claims to rights have proven sometimes odd and intriguing. The Cloning Rights United Front issued statements to the effect that the government has no business interfering with the individual reproductive rights of individuals. Those rights include cloning, they insist. Obviously, the Constitution has no specific clause stating that every individual American has a "right to clone," nor even a "right to reproduce" for that matter. Yet previous rulings have, in fact, upheld a de facto right for individuals to have considerable individual control over reproduction, including some of the cases relating to birth control and abortion discussed earlier.

Now that some states, led by California, have passed legislation to prohibit cloning but to allow stem cell research, we are likely to see cases brought forth to challenge those laws and to seek protections. In the absence of federal legislation or clear legal interpretation, however, the legal questions remain largely in the future. It is to their credit that constitutional law scholars have raised legal questions and called for thoughtful and informed discussion now as a matter of jurisprudence rather than wait until we have to adjudicate violations of particular laws.[22]

The public reaction to cloning is a legacy of the Human Genome Project. So much money and effort, through so many conferences and workshops, has been spent on ethical and legal issues relating to genetics that we risk distorting our understanding of ourselves and our sense of what an individual life is and should be. In considering the genome project and its implications for "reprogenics," as well as cloning and stem cell research and related biological interventions that involve human eggs, women's groups are increasingly pointing to the potential for abuse and exploitation of women. As long as there is a significant market for human eggs and as long as women are allowed to sell eggs, there will be temptations to entice if not to coerce their sale. We must work against such social imbalances without, however, foreclosing the technical development.

What is the value of an egg? Of life? What is a good and valued life? Cloning seems to challenge our very conception of life and what is good, or perhaps it only seems that way because we have not reflected sufficiently on what we value. Certainly some of those commenting on cloning have been all too eager to jump to conclusions in terms of genetic determinants and genomics and to ignore the process of epigenetic differentiation. Their interpretations point to the dangers of rushing to scientific or ethical conclusions in the absence of a richer understanding of the context. The poet T. S. Eliot wrote:

> Where is the Life we have lost in living?
> Where is the wisdom we have lost in knowledge?
> Where is the knowledge we have lost in information?

These lines from "The Rock" (1934) seem so perfectly to capture the lamentations of our modern age that they are widely quoted. A single Internet search netted 477 citations of precisely this passage, as the opening for information technology conferences, for example, or as a clairvoyant Eliot foretelling the information age. Eliot seems to be telling us that we should not get lost in information but should retain a larger sense of life, wisdom, and knowledge. He lamented that perpetual activity had left Man "farther from God" but nearer to "the Dust." Perhaps proximity to dust, and implicitly to nature, is fine. I find it so.

But we must at least ask in what interest we seek knowledge, and to what end we do our living? What is life that we value it; what is a life that we hold it dear; and what might it mean to live a good and valued life? Perhaps our pursuit of an "endless cycle of idea and action, endless invention, endless experiment" will serve us well. But at the very least we need to address these questions.

If we accept the Socratic injunction to reflect and examine what is good, thereby making life more worth living, then one thing our ethical reflections ought to tell us is that we are a long way from defining life or even deciding when it begins. We are also a long way from having intelligent policies or even practices to develop policies that are likely to lead us to wise decision-making. Instead, we rely on the vagaries of reaction and the hubbub of policy making through an endless cycle of reactions to current events, speeches, lobbying, bills, amendments, rejections, rebuttals, and so on.

Policy and Politics

On February 24, 1997, President Clinton had just learned about Dolly along with everybody else. He quickly referred the issue to the NBAC (his National Bioethics Advisory Commission) and asked for a report in ninety days. At the time that schedule seemed both crazy and also sensible, since the public reaction was so strong as to warrant a call for immediate action. Clinton wrote to committee chairman Harold Shapiro: "As you know, it was reported today that researchers have developed techniques to clone sheep. This represents a remarkable scientific discovery, but one that raises important questions. While this technological advance could offer potential benefits in such areas as medical research and agriculture, it also raises serious ethical questions, particularly with respect to the possible use of this technology to clone human embryos." He asked for a "thorough review of the legal and ethical issues associated with the use of this technology, and report back to me within ninety days with recommendations on possible federal actions to prevent its abuse."[23] Just days later, apparently after learning of reports that a rhesus monkey had been cloned and before waiting to hear from his committee, Clinton ordered a ban on the use of federal funds for human cloning.

On Capitol Hill, Democratic Senator Bond of Missouri and Republican Congressman Ehlers of Michigan each introduced bills to ban federal funding for human cloning. Congressman Vernon Ehlers, himself a Ph.D. physicist and one of the few congressmen with any significant scientific training, proposed two straightforward bills that would have outlawed the use of federal funds to clone humans for the purposes of creating a human being. Numerous other bills, none of which actually made it to the floor of the full House or Senate during the 105th Congress of 1997 and 1998, also appeared.

At the time, I had just recently begun working as science advisor to Congressman Matt Salmon of Arizona, and I was astonished that some of the early draft bills were so vaguely worded that they would have outlawed gene cloning or even, in the most extreme case, much of seed reproduction agriculture or even naturally occurring identical twins. Fortunately, in the course of hearings and staff briefings, it became clear that the Congress was not prepared to take action. Several hearings in the House and Senate calmed the most urgent fears that human clones were likely to start popping out like sheep any minute. In addition, what had initially seemed like a clear outcry for legislation and restrictions on this research became complicated. Some conservative groups lobbied not to forestall the possibility that cloning might offer another technology to help otherwise infertile couples have children. Having children is good, they argued; some argued for what they saw as the basic reproductive rights of American citizens. With this lobbying effort, even conservative members of Congress felt it wiser to hold off on legislation that they might later regret.

When the Food and Drug Administration declared in a "Dear Colleague" letter in early 1998 that it had regulatory jurisdiction over "biological products" of the sort involved in cloning and would therefore exercise control over any "clinical research using cloning technology to create a human being," the sense of urgency to develop a special law significantly decreased. This was a huge relief for many people. The FDA imposed oversight without raising prospects of serious reaction by constituents. NBAC's report, released on June 9 and including a call to extend the moratorium on federal funding, surprised no one. The legislation that President Clinton proposed in order to prohibit cloning humans on June 9, embodying NBAC recommendations, did surprise a few people, but more because it found no congressional supporters to introduce the legislation. Ehlers's bill moved through committee but not to the full floor, and the Senate did not act further. It was only with the 107th Congress and only after further scientific innovations that the House actually passed legislation and helped provoke President Bush to make his speech on August 9, 2001. We will return to Bush's political response in the next chapter.

Back to the Science

Once Wilmut and the Roslin team announced Dolly, it seemed inevitable that there would be a rush to clone other animals. There has been. The list of animals successfully cloned continues to grow, along with technical improvements. The list of types of donor cells used also grows, even though most people still do not understand the significant difference between using a differentiated somatic cell and an embryonic cell as the donor.

In some of the clones produced, like Dolly's stablemate Polly, extra genes were spliced into the genetic material for cloning. A gene for human blood-clotting factor was inserted into Polly's genome, but Polly was a sheep, and sheep do not express this gene, so Polly did not develop any human blood-clotting factor. Introducing the gene was a technical achievement, and definitely noteworthy, but it did not make Polly functionally different from other sheep. Similarly, the small transgenic rhesus monkey named ANDi, whose birth was announced in January 2001, had its DNA modified by researchers at the Oregon Regional Primate Research Center in Beaverton. ANDi had an added gene that normally leads to the production of green fluorescent protein in jellyfish. This gene, carried by a virus into unfertilized eggs, did not make the monkey glow green because it was not expressed. Nor was ANDi a product of cloning, but the possibility of merging genetic engineering and cloning was becoming more fact than fiction. The Oregon facility has since announced successful cloning of monkeys from embryonic cell donors rather than from adults.

In public discussions of cloning, advocates occasionally pointed to the advantages for human reproduction, and those in the extremely important but not particularly romantic agricultural world saw its value for breeding more productive livestock. The most common positive use of cloning that caught the larger public imagination, however, was as a tool for propagating endangered species. Maybe Jurassic Park really could exist. In 2000, Robert Lanza of Advanced Cell Technology (ACT) in Worcester, Massachusetts, announced an attempt to clone an endangered ancient ox, the guar. Produced from guar DNA transplanted to a cow egg as the host and implanted in a surrogate cow mother, the guar was delivered at term but died forty-eight hours later from a dysentery that is apparently common in cattle. In October 2001, *Nature Biotechnology* announced the successful cloning of a European mouflon, an endangered species and one of the smallest of wild sheep. The healthy lamb was living at a wildlife center in Italy. Despite the considerable costs involved, for increasing numbers of the public concerned about the loss of biodiversity, cloning technology may offer the best hope of preserving more species and for developing what have been called "DNA banks." There are evolutionary implications, of course, though we are not yet sure what those are.

For less exotic species, the report of a cloned kitten on February 15, 2002, by a Texas A&M laboratory briefly caught the public eye. A peer-reviewed article in *Nature* reported that the experiment produced little CC (for copycat or carbon copy), who remained healthy. The researchers used nuclear transplantation and a surrogate mother to produce eighty-seven eggs, of which only one survived.

More recently, we have heard bold claims that various labs intend to clone humans. On Thanksgiving weekend in 2001, the always publicity-hungry Advanced Cell Technology dramatically announced a successful human clone that had reached the six-cell stage. The spokesperson claimed that the company had achieved two cell divisions with donor cells taken from the cumulus (the structure that nurtures eggs in the ovaries), not later-stage somatic cells. Since a grouping of six cells in the pattern reported is not normal in humans, most scientists expressed skepticism that ACT had actually cloned humans in any legitimate sense.

Scientists immediately denounced the group. Not only were

the data limited and their conclusions suspect, ACT had not submitted the scientific research to an accepted peer-reviewed journal where it could be vetted through the usual critical process. The researchers instead presented their cloning claims in an online magazine, *e-biomed: The Journal of Regenerative Medicine*. They argued that the vastly increased speed of electronic publication is necessary for such an important and fast-moving field. The scientific community, however, remained unpersuaded. Those who choose to issue their scientific claims by such avenues, and by "press release," will face the criticism of colleagues for doing it "wrong." Only if the results can be reproduced and stand the test of time will they be respected as legitimate science.

ACT has been criticized for preying on the public's hopes and fears as well. Whether the scientists at Advanced Cell Technology truly believe that they can successfully clone humans in the near future, or whether they are exaggerating their claims to attract business investment and publicity, they do appear to operate on the assumption that human cloning can be done and that there are at least some circumstances under which it should be done. Kyla Dunn's story on "Cloning Trevor" in the *Atlantic Monthly* shows this side of the cloning business. The poignant case of a family's hopes and the way they run aground on ACT's exaggerated claims make it clear that the cost of hype and insufficient public understanding is high.[24] An Italian team also claims to have cloned human embryos, though we have not seen any results. There will likely be other such claims, since whoever does produce the first healthy cloned human will undoubtedly receive an avalanche of attention, with a mix of approbation and disapproval. We will only know what validity such claims have if and when a child is born. Let us hope that any such child is healthy, as Louise Brown was, and has such delighted and wise parents.

Conclusions—for Now

Many compelling and profound questions arise when the topic of discussion is cloning. What does it mean if we can not only manipulate and design life genetically but also engineer life by putting together pieces from different individuals and create a new being? Does being human require that one begin with sexual reproduction, in the sense of bringing together germ cells from two different individuals of the same generation? Yet an individual cell, even a diploid cell containing a full complement of chromosomes from two parents through sexual reproduction, is not yet fully a life by itself until it develops further. It does not have a nervous system and can do nothing by itself. As Weissman reflected in the *Year of the Genome,* the nervous system "is formed only after the embryo develops in a uterus where two lives remain intertwined until parturition. At term, life begins in a painful act of love."[25]

The questions laid out by cloning for the purpose of producing whole organisms, and in particular whole human organisms, are deep and troubling. But they are not impossible to address. The NBAC report in 1997 focused on what was then known about cloning through the techniques of somatic cell nuclear transfer. The committee concluded that attempting to create a child using this technique is immoral, whether for research or clinical, medical purposes. All members agreed that the techniques were not safe enough and the safety and ethical risks remained too high. Furthermore, society was not ready to move ahead with reproductive cloning without considerably more widespread public discussion. The NBAC therefore offered the clear recommendation to continue the moratorium on the use of federal funding for such research and called for privately funded groups to honor the same constraint. It recommended federal legislation to prohibit such cloning, or attempting to clone, but suggested a "sunset clause" to allow review of the situation in the future. Yet it also noted the importance of not unwittingly outlawing legitimate and useful research using similar techniques on other organisms or other techniques that might be labeled as cloning, including of cell lines. The NBAC sought only to prohibit the creation of an actual human being and to promote thoughtful consideration of the implications of somatic cell nuclear transfer.

A National Academy of Sciences report appeared to draw similar conclusions, but its approach was interestingly different. Note that the NBAC was under considerable pressure to act quickly, since Clinton had asked for a report within ninety days. The NAS report took several years longer, appearing early in 2002. The delay provided the advantage of added perspective: both science and bioethics evolve in light of accumulating knowledge, and the report took that evolution into account. Thus, where the NBAC focused on ethical questions, the NAS panel decided to defer "to others on the fundamental ethical, religious, and societal questions, and presents this report on the scientific and medical aspects to inform the broader debate."[26]

The panel concluded that we should not do human reproductive cloning because it is "dangerous and likely to fail." Like the NBAC, it recommended a ban that would be reviewed within five years. Unlike the NBAC, which did not have the benefit of more experience and more experiments on other organisms, it pointed to concerns raised by the scientific results. The panelists recognized the importance of definitions and took "human reproductive cloning as the placement in a uterus of a human blastocyst derived by the technique that we call nuclear transplantation." Thus, they did not limit the definition to somatic cell transfer, and they also did not consider it reproductive cloning unless the hybrid cloned cell actually developed to the blastocyst stage and was implanted in a uterus. These are both significant distinctions. "In reaching this conclusion," they explained, "we considered the relevant scientific and medical issues, including the record from cloning of other species, and the standard issues that are associated with evaluating all research involving human participants."

Through its various panels and commissions, the United States continues to struggle with its collective values and interpretations of cloning, just as other countries and international bodies are doing. The United Kingdom offers a compromise position, with legislation outlawing cloning of humans for reproductive purposes but explicitly not prohibiting nuclear transplantation to produce cells and tissues either for research or for experimental clinical purposes. Other countries have gone on record with a variety of reactions, with Asia entering the research race most recently.[27] Undoubtedly, the discussions will continue, as will differences of opinion. Yet this public debate is good if, and perhaps really only if, it is informed by a reasoned and reflective appreciation both for the science involved and for the ethical and social interests and concerns.

Hopes and Hypes for Stem Cells

At first the term *cloning* meant producing genetic copies—of cells, genes, or whole organisms. I recall one lively discussion not long after the announcement of the birth of a cloned sheep, Dolly, when I was working as science advisor for my Arizona congressman, Matt Salmon. Salmon, several of his staff members, and I were discussing what cloning really means. The question was whether Salmon should sign on as a sponsor of Congressman Vernon Ehlers's bills designed to prohibit human reproductive cloning, and I was explaining why from Salmon's political point of view he probably should, even though many questions remained. This was spring, 1997, and he asked whether it was possible to use cloning to grow human spare parts. He knew that we were not likely to generate batches of arms, legs, or eyes, but could we perhaps use this technology to generate liver or other valuable tissues or cells? I confidently answered no, that such a prospect was extremely far removed from what the Scottish researchers had done with Dolly, and that Ehlers's bills were carefully worded so as not to prohibit an advance of this sort, anyway. Salmon was concerned, as any thoughtful person should be, that we not be tempted to clone large numbers of fetuses or babies in order to harvest organs from them. Yet he did not want to foreclose the chance that this technique might have unforeseen medical benefits, such as the production of replacement tissues or cells.

In retrospect, I realize now I was too hasty in my response. Given that my own scholarly field is the history of developmental biology, it was all the more foolish of me to sit there explaining why we probably could not culture needed tissues using cloning techniques. Of course, in some literal sense I was right. It was not by using those precise techniques that we would be able to produce replacement tissue. But I was foolish not to have taken the opportunity to explain that science evolves and often twists in unexpected ways, that we should expect to make new discoveries, and that we should think about the range of possibilities in order to be in the best strategic position to react intelligently rather than only instinctively. History teaches us that we should expect to be surprised. Our policies should provide flexibility and allow for change in response to new discoveries, new opportunities, and new concerns.

In 1998, two scientists reported their surprising results with stem cells (see section on "Stem Cell Science," below). They raised the possibility that some forms of cloning might have medical value, even if we choose not to reproduce complete normal human beings. Combining cloning and stem cell technologies might make it possible to generate replacement tissues with significant therapeutic value. This technique was soon labeled "therapeutic cloning," though many scientists since have tried to distance themselves from that term. It is worth understanding precisely what is at issue and reflecting on how we got to that surprising point.

Early Stem Cell History

When *Science* declared stem cells the "breakthrough of the year" in 1999, it was, indeed, an exciting time for stem cell research. Yet the story did not begin with the dramatic announcements in 1998. Scientists had understood for decades that stem cells are the reason that bone marrow transplants help some leukemia patients, so the concept of stem cells was not entirely new. What changed—what had changed, too, with the cloning of Dolly—were our preconceptions. In this case, our ideas changed about the limits of stem cell capabilities and of how differentiation occurs, epigenetically, during development.

Already by the seventeenth century, blood transfusions had been carried out successfully from animals to humans, though it was only in 1795 that Philip Syng Physick reported having transferred blood from one human to another. In the twentieth century transfusion became common, as scientists worked out ways to control immune responses, discovered the different human blood types and systems for matching them, improved anticlotting control, and added other incremental advances. Even though it is now more routine, transfusion is never risk-free and the blood supply does not insure that a donor is always available with the right kind of blood and enough of it. Especially for treating conditions like leukemia or some cancers requiring radiation treatments, the need for large quantities of new blood raised questions about whether it might be possible to find ways to cause the body to produce more of its own supply.

By the middle of the twentieth century, researchers began to think that they might combine the accumulating knowledge about those special, apparently not yet differentiated "stem cells" with their understanding of physiology. Perhaps they could make use of stem cells—particularly hematopoietic stem cells from the bone marrow, which normally give rise to blood cells—to cause the body to produce more of its own blood cells.

Experiments began in earnest in the late 1950s in France, after a radiation accident destroyed patients' healthy blood cells and produced types of leukemia. As the National Marrow Donor Program history explains, in 1958 Jean Dausset discovered the first human protein that allows the body's immune system to distinguish its own cells from foreign cells.[1] The human leukocyte antigen (HLA) in the donor blood cells elicits an immune response from the recipient, whose body rejects the foreign cells, including ones that could be extremely beneficial. The destruction of foreign cells makes tremendous sense on evolutionary grounds, since most invasive entities are harmful—think of bacteria and viruses. Usually the body benefits from an ability to recognize "non-self" cells and destroy them. The challenge for medical science, then, was to override the evolutionary safety systems.

By the 1960s, researchers had worked out enough details about the system to transplant bone marrow successfully from a sibling into a child with immunodeficiency disease (the so-called "bubble boy syndrome"). Further modifications in the method brought the first successful marrow transplant to an unrelated patient in 1973. By the early 1980s, the approach worked so well that it was the best hope for many leukemia patients. Unfortunately, it was (and is) still necessary to find a close bone marrow match, and that was difficult. The parents of a young woman who had received a successful transplant but had died later from a recurrence of the disease decided to help others and began what has become an international bone marrow donor

WHOSE VIEW OF LIFE?

network. In 1984, Congress passed the National Organ Transplant Act to support the system of identifying donors and potential recipients of transplantations in the United States and beyond.

The public knows of the donor network because of publicity about organ and bone marrow transplantation and the need for donors. We read about the plight of over 80,000 patients who were listed as potential organ recipients by 2002, thousands of whom will die every year because of lack of available organs. Fortunately, the bone marrow network seems to reach a higher percentage of needy patients, though many still die without treatment. By 1996, the marrow network had arranged 5,000 transplants a year and maintained over two million potential donors in its data bank. What the public did not understand before 1998 is that bone marrow transplanting succeeds because of stem cells.

The identification of stem cells as special cells capable of giving rise to a variety of other specialized cells was made in 1896 by E. B. Wilson. Wilson's friend William Sedgwick had used the term a decade before for the first time, applied to plants, but Wilson gave it new meaning. Writing about the roundworm *Ascaris*, he noted that some cells specialize but that some "may be called the *stem-cell*" as they retain all their chromosome material and do not divide into pieces as other cells do.[2] It was the 1950s when biologists recognized, from their experience with marrow transplantation, the full significance of the apparent capacity of these cells to serve as generalized starting points for other types of cells. Then animal stem cell research began in earnest.

The public did not hear about the stem cell studies on mice and mammals that started in the 1960s and 1970s. Yet it is this work that led to the discoveries announced in 1998. By the late

1960s, Cambridge physiologist Robert Edwards noted that when he began studying the cells from the pre-implantation blastocyst (the stage just before implantation in the uterus and just before differentiation of cell types), he achieved astonishing results. Further, Richard Gardner showed that these embryonic stem cells could be inserted into cells inside the mouse blastocoel (the hollow area inside the developing embryo), where the two types of cells would mix together and produce organs and parts for the resulting chimera or hybrid. Gardner's first mouse hybrid was made up of cells taken from different animals and had a mixed coat color that made it instantly recognizable as a blend.[3]

The resulting transgenic mice raised questions about what might be possible with further research. Excited about the work, Edwards nonetheless had to "place the stem-cell projects on the back burner" because of the demands of the *in vitro* fertilization work that he had begun with Patrick Steptoe and that led, eventually, to the birth of Louise Brown and an entire fertility industry. Edwards notes that, far from being unknown among scientists at the time, "'ES' cells [embryonic stem cells, those derived from embryos] became a familiar term, reflecting their immense potential."[4] Indeed, the term was closely linked with studies of the early stages of human development as biologists sought to learn under what conditions fertilized eggs would grow in a petri dish until they could be transferred to a woman, undergo implantation, and begin their normal gestation inside the womb.

Researchers also made significant advances with mouse pluripotent embryonic stem cells through the 1980s. Whereas a *totipotent* cell can become an entire organism, including all the different cell types, a *pluripotent* cell has the capacity to become any one (but not all) of the cell types that make up the body. *Multipotent* cells are a little more specialized and can become one of several, but not of any, cell types.

It is, of course, difficult to study one cell isolated all by itself, since most research destroys the cell. An approach that has proven successful since the early twentieth century is cell culture. A cell or group of cells can be grown in a specially formulated culture medium, in which they may divide over and over again. A pure cell line produces many generations of cells just like the founder cell or cells and in theory can be "immortal" and perpetuate itself forever. Such cell lines can provide the many cells needed for research.

By the 1990s, several human cell lines had proven their pluripotency, or at least their multipotent ability to become more than one kind of cell. It was now clear that the cells at the blastocyst stage (before implantation in the uterus) and occasionally cells from other stages are not yet irreversibly differentiated and therefore offer promise for further research. Yet the public, and even most scientists, did not fully realize just how immense that potential was until two researchers gave hints what might be done with human stem cell lines in 1998. The time was ripe for an explosion of new discoveries and clinical innovations—if politics and bioethical concerns did not stop them.

Developmental Stages

Before getting to 1998, let us review some of the important terminology of human developmental biology. Words from the specialist jargon that drives so many students crazy in science classes suddenly sprang onto the front pages of the leading newspapers and out of the mouths of television newscasters. We need to know what these words actually mean because they refer to new concepts and content. The issues raised are too important to be left to scientists alone or to let the general public react intuitively, ignoring the scientific context. It is worth reviewing where we were in understanding human development as of early 1998.

The earliest stages of an ordinary life are triggered when boy meets girl, as a sperm cell meets an egg. The nucleus of each cell has one set of chromosomes, so that when the two nuclei join the fertilized egg has the full complement of nuclear chromosomes (two of each type), with their DNA and the genes that contribute to making up the human being. The fertilized egg begins to divide, directed by the internal mechanism of mitosis and guided by the structures in the egg's cytoplasm (that is, the material that is not in the nucleus or the outside coating of the egg). The importance of this cytoplasm is often ignored by genetic determinists but not by developmental biologists, who realize that genes cannot do anything without the apparatus of the cell to guide them.[5] Most fertilized eggs never develop at all or survive past a few cell divisions, because of "defective" genes or for other biochemical or mechanical reasons. We understand little about all the factors that can cause developmental problems, though researchers have made tremendous progress in interpreting the relative genetic and developmental, or epigenetic, contributions in animals other than humans. Until the coming of *in vitro* fertilization, researchers were studying frogs and other species, not humans. Culturing human eggs in a dish and then throwing them away seemed unacceptable, until there was a medical reason to do so. With the impetus of fertility medicine, however, knowledge began to accumulate quickly.

First the fertilized human egg divides into two more or less equal cells, then four, then eight. Up to this point, it remains about the same size and just divides into smaller units. Also up to the eight-cell stage, each cell is essentially equivalent (see Figure 8). We could divide the eight cells of a human embryo into eight separate dishes and culture them so that they will all grow.

WHOSE VIEW OF LIFE?

We can remove a cell or several, and the remainder can still adjust and develop normally. In fact, some human genetic testing does just this and the remaining six or seven cells produce apparently perfectly normal babies. For eight cells to separate and each to produce a baby would certainly be rare, though it is theoretically possible; no octuplets have lived a full life, though septuplets have. With the next cell division, by the sixteen-cell stage, some differentiation does begin to occur: sixteen-tuplets are apparently not possible. In theory, it should also be possible to take eight cells from each of eight different dividing eggs and then put them together to make up one new hybrid embryo. Indeed, evidence is accumulating that cells do come together from different eggs under extremely rare circumstances, even in humans.[6] The point is that all the cells to the eight-cell stage apparently retain complete "totipotency." Each one, just like the fertilized egg cell, has the ability to become a whole normal living human being.

This does not mean that when they are all together the individual cells do not have any differences, but they retain the capacity to "regulate" and reshape the whole if needed. They can be separated, yet when they are together signaling among them coordinates their divisions and activities during the early stages and thereafter. This early cell division takes us roughly through the third day. Most significantly, the capacity of individual cells separated from the others at the early stage to develop as a whole raises problems for those preformationists who see a life as beginning fully at conception. How can a preformed life regenerate missing parts or otherwise adjust to changes, epigenesists ask?

The proliferation of cells continues for several more cell divisions, though the divisions become less regular and symmetrical. By day four, the cells begin separating and migrating so that

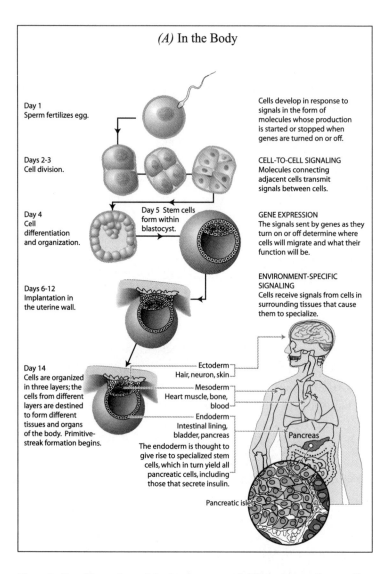

(A) In the Body

Day 1
Sperm fertilizes egg.

Days 2-3
Cell division.

Day 4
Cell differentiation and organization.

Day 5 Stem cells form within blastocyst.

Days 6-12
Implantation in the uterine wall.

Day 14
Cells are organized in three layers; the cells from different layers are destined to form different tissues and organs of the body. Primitive-streak formation begins.

Cells develop in response to signals in the form of molecules whose production is started or stopped when genes are turned on or off.

CELL-TO-CELL SIGNALING
Molecules connecting adjacent cells transmit signals between cells.

GENE EXPRESSION
The signals sent by genes as they turn on or off determine where cells will migrate and what their function will be.

ENVIRONMENT-SPECIFIC SIGNALING
Cells receive signals from cells in surrounding tissues that cause them to specialize.

Ectoderm
Hair, neuron, skin

Mesoderm
Heart muscle, bone, blood

Endoderm
Intestinal lining, bladder, pancreas

The endoderm is thought to give rise to specialized stem cells, which in turn yield all pancreatic cells, including those that secrete insulin.

Pancreas

Pancreatic islets

Figure 8. Two illustrations of the development and differentiation of stem cells: *(A)* in the body and *(B)* in the laboratory. The example focuses on pancreatic cells that produce insulin.

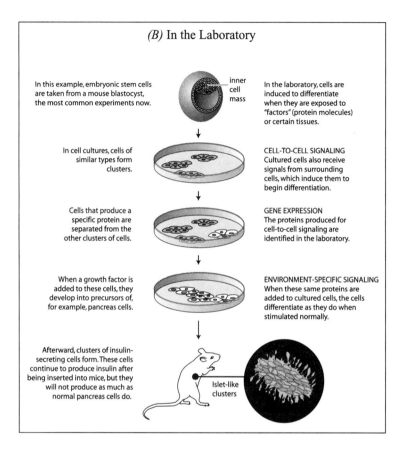

(B) In the Laboratory

In this example, embryonic stem cells are taken from a mouse blastocyst, the most common experiments now.

inner cell mass

In the laboratory, cells are induced to differentiate when they are exposed to "factors" (protein molecules) or certain tissues.

In cell cultures, cells of similar types form clusters.

CELL-TO-CELL SIGNALING
Cultured cells also receive signals from surrounding cells, which induce them to begin differentiation.

Cells that produce a specific protein are separated from the other clusters of cells.

GENE EXPRESSION
The proteins produced for cell-to-cell signaling are identified in the laboratory.

When a growth factor is added to these cells, they develop into precursors of, for example, pancreas cells.

ENVIRONMENT-SPECIFIC SIGNALING
When these same proteins are added to cultured cells, the cells differentiate as they do when stimulated normally.

Afterward, clusters of insulin-secreting cells form. These cells continue to produce insulin after being inserted into mice, but they will not produce as much as normal pancreas cells do.

Islet-like clusters

a layer of cells accumulates around the outside of the sphere. Other cells collect inside and continue dividing. The outer layer will become the placenta and the tissue that supports the development of the fetus, while the inner mass of cells becomes the body. By the fifth and sixth days the hollow ball of cells, called the "blastocyst," apparently has the greatest value for therapeutic use. At this point, those internal cells are not yet differentiated but they are fully ready to become so, since each cell is still pluripotent. With "immortal" cell lines from these cells, researchers can give rise to more stem cells through a number of generations and also, if stimulated in various ways, to differentiated cells of different types. Too little research has been done on human embryonic stem cells, however, to be sure exactly at what point these internal cells begin to become "channeled" or limited in their capacity to become some kinds of cells and not others. One reason we would benefit from considerably more research on embryonic stem cells is that, by knowing exactly how they work, we would have the greatest possible chance of developing effective clinical treatments using stem cells.

The blastocyst then moves to the mother's uterine wall and, by the ninth or tenth day in humans, it is completely implanted there. It continues differentiating and begins exchanging nutrients and waste with the mother. The implantation stage marks a significant change. Up to this point, as Edwards had demonstrated in his *in vitro* research and clinical successes since the 1970s, it is perfectly possible to culture the egg in a glass dish. For this reason, some people have called the dividing cells a "pre-embryo" or a "pre-implantation embryo" until this stage occurs. Developmental biologists used to call the fertilized egg an "embryo" just as soon as it started dividing, however, just because that was easier. There seemed to be no reason to make further distinctions. Until recently, exactly how to define the term *embryo* has made little difference for biologists, since there is no natural distinction between an embryo and an almost-embryo or a just-finished-being-an-embryo. The distinction is a matter of definition, or convention. What really matters are the biological details as the egg develops, and as biologists over a century ago began learning more about these details they assigned names to the developmental stages. *Morula, blastula, gastrula,* and other designations are the descriptive terms that came into use because they make sense for embryologists and developmental biologists. Now, for political reasons, the definitions matter.

What is that fertilized egg sitting there in the fertility clinic dish? Is it an embryo? A pre-embryo? Is it a life? It is clearly the material beginning for processes that will eventually lead to a formed human, if all goes well. It has not really begun to differentiate into different kinds of cells. It is not yet fully a life in the traditional Aristotelian, Jewish, medieval Catholic, Muslim, or conventional secular sense. Leading scientists Bert Vogelstein (as a Howard Hughes Medical Institute investigator at Johns Hopkins University's Medical School and chair of the National Academy committee reporting on stem cells), Bruce Alberts (as president of the National Academy of Sciences), and Kenneth Shine (as then-president of the Institute of Medicine) agree. Commenting on "cells growing in Petri dishes," they said: "The public needs to understand that there is a huge difference between such cells and an actual human being."[7]

For many biologists, biologically meaningful life begins only after the biologically significant stage called "gastrulation." As embryologist Lewis Wolpert put it in his *Triumph of the Embryo,* gastrulation is "truly the most important event in your life." Or as neuroscientist Michael Gazzaniga stated at a meeting of President Bush's bioethics council: "It is a truism that the blastocyst has the potential to be a human being. Yet at that stage of development it is simply a clump of cells . . . An analogy might be what one sees when walking into a Home Depot. There are the parts and potential for at least 30 homes. But if there is a fire at Home Depot, the headline isn't 30 homes burn down. It's Home Depot burns down."[8]

Soon after gastrulation comes development of the "primitive streak." This stage occurs by the fourteenth day, and it is marked by the appearance of differentiated cells that apparently have some capacity to experience sensation. This capacity to feel offers a point of difference that many believe is the meaningful beginning of a real life. By the primitive-streak stage, to pursue Gazzaniga's metaphor a bit further, the embryo has begun to represent the early foundations of a house rather than just assorted boards and nails. It is not yet a house, however, or even the full frame for one.

Others point to later stages as the most important, such as the twenty-second day, when the heart begins to beat. By the fortieth day, most of the body parts are recognizable, at least in primitive form, and by eight weeks the embryo has become what is labeled a "fetus"; it looks like a human being and has all the functioning parts and organs at least in rough form. What should be clear is that there are many biologically significant steps along the way toward a life. It is not the case that a life begins at conception in the sense that it is recognizable, or that it has all its parts, or that it is functioning in any significant way other than cells dividing into other cells that are just like them. Implantation, primitive streak, forty days, eight weeks: each has biological significance, as does every other point along the way. Biology alone is not going to give us the wisdom we need to guide our political and social decisions about when a life begins any more than the Pope's decree can do so.

Stem Cell Science

Next, in 1998, two researchers challenged what we thought we knew about human development. Their discoveries about stem cell abilities followed earlier research, and their work would surely have led to further important publicly funded and publicly disclosed discoveries had anti-abortion politics not forced the research into the private and for-profit sectors. James Thomson, working at the University of Wisconsin and funded by Geron Corporation since federal funding was not available for human embryo research, published a paper in the November 6 issue of *Science*.[9] He explained that he and his collaborators had successfully cultured human stem cells. Because of his earlier work on primates, he recognized some cells as characteristically stem cells and pursued research on them in a way that many researchers would not have known to do. Shortly after that announcement, Johns Hopkins University's John Gearhart, also funded by Geron, announced that his team had successfully isolated and cultured human embryonic stem cells.[10]

In each case, the earliest findings were more suggestive than definitive. Both studies demonstrated that human stem cells could be sustained in culture through many generations and still show all indications of differentiation after that. Much research remained to be done, of course. What caught the public attention was, first, that the cells seemed to have an almost magical power to become whatever we want them to be. Second, the public noticed, because the press and the political lobbies made very sure to tell them, the cells came from contested sources.

Both Thomson and Gearhart remained squeaky clean in terms of the legality of their research. Since neither used public funding to derive or to manipulate the human cells, they did not violate any laws nor raise any question of doing so. What did raise questions, however, is what the anti-abortion lobby saw as unethical—even though perfectly legal—practices in creating these cell lines. Thomson had obtained his cell lines from embryos that had been frozen in fertility clinics and were destined for disposal. He made very sure to obtain all appropriate permissions; in fact he went beyond what was legally required. Gearhart had used cells taken from fetuses that also were donated and obtained with all proper permissions. It was all perfectly legal. That is precisely what bothered the anti-abortionists.

Hopes and Fears

Even where their opinions are in the minority, the anti-abortion lobby in the United States holds such strong influence with the public and with recent Congresses that when it speaks, it is heard. With the announcement of stem cell research, it spoke—loudly. Calls rose for federal legislation to prohibit research on stem cells on the grounds that it was done on embryonic or fetal cells.

Very quickly, the few researchers who had been exploring adult stem cells found themselves thrust into the spotlight because their work looked like an acceptable alternative line of research. Biologist Margaret Goodell was one of those who found herself surprisingly and suddenly in the public eye. She was called on to testify before congressional hearings and was confronted by journalists. As she told a reporter for *Science*, she was concerned that people not rush to conclusions. Yes, some of the adult stem cells on which she did research might have some potential to develop into various cell types, but little work had been done on them yet and there was little in the way of concrete results to offer. They are not the same as embryonic stem cells (ES cells), she noted confidently. "Embryonic stem cells have great potential. The last thing we should do is restrict research." Rather, she said, we should study both embryonic and adult stem cells, in the hopes of developing both understanding and potential medical applications.[11] This is especially true given that the accumulated evidence strongly supports the conclusion of the NIH's massive report on stem cells that the adult cells are different from, and do not have the therapeutic potential of, embryonic stem cells. Indeed, the cells that Goodell had thought were stem cells for muscle turned out not to be so.[12]

During 1999, many researchers began to apply to human stem cells the knowledge gained from decades of studying mice. Because of politics and fears that their funding might disappear suddenly, however, most researchers who might have moved into this area held back. By the time the NIH decided to move ahead and fund limited stem cell research late in 2000, the worries had grown sufficiently that only two proposals were submitted and the program was halted before it began. Reasonably enough, researchers hesitated to invest much effort or to lure graduate students or postdoctoral fellows into research that might at any moment be shut down by federal legislation. Researchers waited, or they looked to private funding.

Advanced Cell Technology took up stem cell research. Just after Thanksgiving, 2001, the company announced a derivation of stem cells from unfertilized eggs (through parthenogenesis). The public was not sure what to make of this announcement, though as Jacques Loeb had shown over a century before, parthenogenetic development calls into question our fundamental assumptions about what a life is and what it is capable of doing in development.

This research by ACT was done in primates, not humans, yet the techniques are presumably transportable. Without fertilizing eggs, the group of researchers from ACT and elsewhere derived cell lines that they had cultured through many generations, for ten months already. The implications were exciting. As in the 1890s, when Loeb achieved artificial parthenogenesis in sea urchins, pundits asked who needs the male. Now the question was who needs the normal embryo? If the egg is not fertilized but still can be caused to multiply and divide and generate lines of cells that behave like pluripotent stem cells, then perhaps this line of research could avoid the ethical complaints against using fertilized eggs that might become humans. ACT's one-page report in *Science* in February 2002 left open many questions, but it raised enough possibilities that some opponents of using embryos or fetal tissue acknowledged that this technique might make an ethical difference.[13]

Further promising research with neural stem cells by several laboratories offered still more surprises, such as the possibility that not all our brain cells are differentiated and fixed by the time we are adults. Perhaps neural cells can be grown in culture and added to patients to promote brain function. This seemed to be true in mice, though only with very preliminary results and more questions than answers. Already by December 18, 2001, researchers had made sufficient progress and the issues had become sufficiently focused that the *New York Times* devoted an entire thoughtful and stimulating Tuesday science section to stem cell research and what it might mean. Because so little research has been done with humans, we are likely to continue to be confronted with surprises, new questions and new possibilities.

Clinical Promises, "Therapeutic Cloning," and Regenerative Medicine

One of the major impediments to using cultured cells for therapies is rejection by the immune system. If we could only use the patient's own cells for the culture, the body would not reject the transplanted cells, which could then replace the function or the product that is missing. This is the driving assumption behind the combination of stem cell research and cloning that was called "therapeutic cloning."

Since the adult patient is obviously no longer a blastocyst or a fetus, we cannot obtain the embryonic stem cells that are most likely to remain undifferentiated and hence pluripotent. Therefore, why not bring cloning technology to stem cell research? The idea is to clone a somatic cell from the patient, allow it to divide to the blastocyst or stem cell stage, then culture it to produce cells of the desired type with the patient's genome. In addition to avoiding rejection of foreign cells, this method would avoid the significant problem of gene therapy—namely, the need to inject the "good" gene into a patient's cells by means of a vector, often a virus, which presents its own risks. In the case of stem cell therapies, medical researchers believe the risk would be reduced because the stem cells come from the patient's own DNA and cell lines. "Perfect. Let's get busy"—that would be a reasonable response to the prospect of helping real people in need. Yet this reaction, too, may run aground on the absolutist views of anti-abortion politics and the opposition to any embryo research. Most people at least see that there are questions to be taken seriously.

Researchers at Advanced Cell Technology provide an alternative view, with their willingness to push ahead and try therapeutic cloning without feeling any ethical constraints. An *Atlantic Monthly* story details ACT's efforts to clone a young boy called "Trevor Ross." Kyla Dunn, who followed the story for six months, makes clear that the scientists at ACT are serious about their work and believe that they can, in fact, successfully develop cloned embryos from which stem cells can be used for therapeutic purposes. They have not succeeded yet, however.

The report shows how many obstacles there are to realizing the hopes—and hypes—of this technology. Just getting enough freshly donated eggs is a significant challenge, especially when the laboratory depends on "extra" eggs that are not needed by fertility clinics. Ethicists also worry that having to obtain the probably thousands of eggs needed to carry out the research even to reach the embryo stage will lead to the "commodification" of women. Yet this is not really different from what fertility clinics do already, except that in those cases the young women donating eggs believe they are selling their eggs to help infertile couples have healthy babies. With research into "therapeutic cloning," in the early stages donors are simply helping with the research that may—or may not—someday help save someone's life or cure debilitating and degenerative diseases.

That was the hope with Trevor, who was afflicted with a rare and degenerative genetic disease. If researchers at ACT could develop a way to insert cells that could perform normally into the boy's body, the cells would fill the physiological gap left by the genetic defect and solve Trevor's problem. They could also make news, generate huge publicity for their company, and provide tremendous incentives for Congress not to outlaw such promising lines of research. Nonetheless, the story remains focused on Trevor. The parents are eager to pursue all options and all threads of hope, as parents typically are. The researchers try and try but meet with heartbreaking failures.

Without public funding, the story argued, they do not have enough eggs to work with because procuring eggs is so expensive and medical insurance does not pay for research. Scientists do not have enough opportunities to try. Research takes many trials and many errors before achieving success, if success is even possible. Sometimes the expected fails, but something else works. It often takes many people trying many things in many different labs, exploring as many options as possible, to bring success. That is not happening without federal funding; meanwhile, private investors remain cautious while they wait to determine whether Congress will enable, outlaw, or fail to act on stem cell research and therapeutic cloning.

This particular *Atlantic Monthly* account about Trevor is carefully written to get at the mix of contradictory messages involved. The scientists are presented as probably overly optimistic and also well-meaning: although they are not purposely misleading the poor parents, they may be doing so unintentionally. The parents are acting responsibly in trying everything to help their son, but they may be gullible in placing all their hope in an experimental process. It has never worked, so why should it now? But there had never before been a baby born as a result of *in vitro* fertilization before Louise Brown. If there were no courageous patients and no hopeful supporters, we would not make progress. The challenge is to achieve a delicate balance between realism and hope, protection and freedom.

Michael West, ACT's president and CEO, feels clear on the moral grounding. "You can be as pro-life as you can get," he insists, "but you can't say that making and destroying a pre-implantation embryo is the destruction of a human. Because it isn't. If it was a human life, I wouldn't touch it. Absolutely not." He is clear that "A human individual does not begin at conception. It begins at primitive-streak formation."[14] For West and other researchers, it is a travesty that sentiment against reproductive cloning has spilled over to research into stem cell development and therapeutic cloning. He clearly feels that the research must go forward.

Other scientists are equally eager to see the science progress even while recognizing the social and moral concerns. A number of biology's luminaries have written editorials and letters calling for us to continue research in the interest of pursuing the possible benefits. Nobelist Paul Berg, a molecular biologist recently retired from Stanford who was a central player in the recombinant DNA debates in the 1970s, wrote that we need strong research to continue in order to discover the clinical applications of stem cell research; neither rhetoric alone nor political decisions can answer those important questions. For policy, we need a clear definition and reconceptualization of life.

Developmental biologist Irving Weissman and the Nobel Prize–winning president of Caltech, David Baltimore, noted in *Science* not that research should continue no matter what, but that we should avoid the "devastating" cost of allowing religious and moral interests to overrule the possibilities of clinical value for many. These men recognize that scientists alone should not make all ethical decisions about their work and that governments have appropriate roles to play in establishing policies governing science. Nonetheless, "in making those policies, the state should minimize purely political considerations and be mindful of the separation of church and state. The wrong action here could close the door to an important avenue of scientific and clinical discovery. The state should not be the barrier to the translation of these potentially revolutionary therapeutic opportunities into real medical advances."[15]

Weissman headed a panel for the National Academy of Sciences on stem cell research, and he has done research and established companies that carry out adult stem cell studies. In a substantive editorial letter in the *New England Journal of Medicine,* Weissman called for more study. At present, he noted, we still know little about the therapeutic realities of stem cell science. Probably some of the so-called stem cells will prove not to be truly pluripotent and will not therefore have the wide range of applicability that the public hopes. Yet we cannot know the realities without doing the research. The existing cell lines, whose use President Bush allowed on August 9, 2001, do not provide enough research material.

Furthermore, Weissman worried that the Congress would pass legislation governing research on human embryonic stem cell lines, and he urged us to recall that twenty-five years ago we seriously considered a legislative ban on recombinant DNA research. Yet, "hundreds of thousands of people are alive or healthier because of the use of recombinant DNA to produce insulin, erythropoietin, granulocyte colony-stimulating factor, interferons, and other important therapeutic recombinant molecules." "I believe," Weissman continued, that research with embryonic stem cells "will be as valuable to medicine as recombinant DNA has proved to be."[16]

Bioethicist Arthur Caplan used stronger language. Labeling those opposed to therapeutic cloning as the "anti-cloners," he criticized them for blocking worthy research and for obfuscating central issues about the costs and benefits of serving diverse, coexisting interests. "Half a Loaf Is Not Good Enough," Caplan asserted late in 2001 in response to the president's decision to allow limited use of existing cell lines. "The Bush compromise is no compromise," he concluded, but really "an absolute ban on embryonic stem cell research" since it allows too little to do any real good and offers too little incentive for researchers to take on expensive research that might be banned at any time. Bush's response was "pathetic." "The American people do not support the view that it is wrong to use embryos that already exist but are going to be destroyed for scientific research as long as there is consent for their use."[17]

Nine months later, Caplan made the case for allowing research even more clearly and strongly. He perfectly well understands the concerns of critics who abhor the creation of embryos for the use of their parts. Most would agree that we do not want to use any human person simply for instrumental reasons, as philosophers put it, or in other words simply as means for another person's ends, however good those may be. That is quite generally regarded as unethical. Yet, as Caplan argues persuasively and powerfully, there are no compelling secular grounds on which to consider those frozen embryos stored in fertility clinics as persons or even as potential persons. "As a simple fact of science, embryos that reside in dishes are going nowhere." "The debate over human cloning and stem cell research has not been one of this nation's finest moral hours. Pseudoscience, ideology and plain fearmongering have been much in evidence. If the discussions were merely academic, this would be merely unfortunate. They are not. The flimsy case against cloning for cures is being brought to the White House, the Senate and the American people as if the opponents held the moral high ground. They don't. The sick and the dying do."[18] Whether we agree or not, the stakes are simply too high not to be very clear and thoughtful about what we are doing and why we are doing it, as well as what we can learn from history about the current controversy.

Many, many people agree with Caplan and have high hopes for what has come to be called "regenerative medicine." This includes the prospects for regenerating body parts and tissues and brings with it the prospects of longer, though surely not eternal, life. Already, we can replace mechanical parts like arms or legs with prosthetic substitutes, or hips and knees with synthetic internal replacement parts. We can transplant organs, including some essential organs without which life cannot continue. These medical solutions all involve replacements.

What if we could find a way to trick the body into acting normally? What if we could cause it to make its own replacements through a sort of regenerative medicine? That is the point of

WHOSE VIEW OF LIFE?

gene therapy: add a gene that can cause production of a missing protein and cure the disease resulting from the absence of that protein. Try replacing the body's capacity to make new liver cells, which we have already begun to do by transplanting smaller and smaller pieces of liver. Add a bit of neural tissue and watch it replace missing or degenerating nerve cells. What if we could transplant stem cells and cause them to develop spinal nerves, or neurons in the brain, or cells whose degeneration causes paralysis, Alzheimer's, or Parkinson's disease. Wouldn't it be wonderful if we could regenerate the missing cells and therefore recover the failed function? That is the hope of regenerative medicine.[19]

Clearly, we need to understand much more about what happens during normal development in order to use that knowledge to guide regeneration. About a century ago, just as Loeb was exploring the prospects for "engineering life" and as Morgan was beginning his studies of genetics, researchers were also enthusiastically studying regeneration. Some animals have a natural capacity to regenerate parts, and Morgan among many others asked why and how they do that. We are only now returning to that study, with a new emphasis, and realizing that study of regeneration can inform our understanding of development, which in turn can inform our pursuit of techniques for artificial regeneration.

That is what we are trying to do now, to trick the body into regenerating cells and tissues where it normally would not be able to do so. For this we need more research. Specifically, we need to study humans to understand humans. Since it is embryonic cells that are developing and beginning to differentiate, it makes sense that we will benefit from studying human embryonic cells and tissues. We have to decide as a society how to balance the interests of all the potential patients who would benefit from the therapies derived with this knowledge, on the one hand, and the interests of the embryos or their protectors, on the other hand.

Who's the Expert?

Who is the expert who can guide us when beliefs are in sharp conflict and considerable public interest is at stake? Questions about expertise bear on most issues relating to science, of course, and we have developed protocols to certify who will count as an expert and how that expertise will be used. For example, the Federal Judicial Center publishes a *Reference Manual on Scientific Evidence,* a thick red book nearly 650 pages long. And this is only one of the guides that judges need to address all the questions they face.

The conditions for bringing science into the courtroom changed suddenly for judges in 1993 with the ruling in *A. Daubert v. Merrell Dow Pharmaceuticals.* The case concerned the admissibility of scientific evidence. Many courts had previously relied on an earlier test of admissibility, a test established by the 1923 *Frye* case relating to the use of polygraph results. There the court determined that expert witnesses should be allowed to give evidence if, and only if, that evidence comes from science that is sufficiently well established to have gained "general acceptance" in its field. *Daubert* allowed judges to determine whether expert testimony achieves a "reasonable scientific standard." Congress had passed legislation in 1975 providing "Rules of Evidence" that were similar in most ways to *Frye* guidelines but did not demand "general acceptance" of the scientific approach. Courts in some states continue to follow "Frye rules," while others have modified their approach and determined that judges have considerably more latitude and more responsibility to determine what science should be allowed into the courtroom.

Daubert opened the way for some highly contested science to appear in the courtroom and for other science to be ruled out, and the case left open what should count as expertise even in order to determine what expertise to allow. *Daubert* in effect made judges the gatekeepers as to the "relevancy" and "reliability" of "sound science" while providing no clear guidelines on how they are to play that role. As one judge put it, "I think we've lost our way, at least this judge has lost his way, in terms of what constitutes reliable testimony." He warned that "A judge potentially is playing a much more developed role in deciding who's going to win and who's going to lose. And we're deciding that, frankly, based on standards that I don't entirely understand."[20] The Federal Judicial Center, American Bar Association, state court associations, and other groups that provide educational programs for judges have argued that we need better scientific advice and a clearer sense of what should count as "sound science" at all levels in this country.[21]

The judicial system is not alone in its struggle to evaluate scientific expertise. Congress used to be able to appeal to the Office of Technology Assessment (OTA) for advice about science and technology, but that office was closed on September 29, 1995, because congressional leadership at that time, seeking a publicly visible way to save money, did not consider it essential. As a current website reflecting on the OTA's legacy explains, the office "provided Congressional members and committees with objective and authoritative analysis of the complex scientific and technical issues of the late 20th century. It was a leader in practicing and encouraging delivery of public services in innovative and inexpensive ways, including distribution of government documents through electronic publishing."[22] Many scholars and congressmen lament the passing of the OTA and offer periodic calls for reinstituting the research group, which was dedicated to providing nonpartisan scientific assessment as objectively as possible.

The Congressional Research Service does provide information to Congress, but it is overwhelmed with requests about all aspects of government and cannot begin to provide all the wide range of scientific expertise that congressional staff and members of Congress need. The Office of Science and Technology Policy for the White House was a reasonably strong and effective source of information and guidance, but the office and its advisor have been demoted in stature under the current President Bush. The National Academy of Sciences, which is historically government-authorized but not directly government-funded nor a government agency, can oversee studies that are commissioned and paid for by Congress, but it does not have the resources to take on many studies of its own.

Funding Stem Cell Science

When President Clinton heard about human stem cell research, he immediately asked the National Bioethics Advisory Commission (NBAC) to examine the ethical issues and policy implications. The commission received the charge on November 14, 1998, and completed its report by September 7, 1999. This time, the public reaction and the commission's interpretation were considerably more complicated. Cloning to reproduce humans had had relatively few advocates.[23] In contrast, if embryonic stem cell research or perhaps therapeutic cloning could lead to successful clinical treatments, many people would want the new therapies. Nearly everybody knows someone who could benefit—someone with a degenerative disease like Parkinson's or Alzheimer's, a spinal cord injury, muscular degeneration, or liver disease—so the issue becomes a very personal matter. As a result, the balance sheet of costs and benefits appears significantly different for stem cell research. Lobbying to support not only allowing such research but also enabling it through significant public funding began quickly, alongside lobbying from those who hate the idea of any embryo research at all, no matter what.

There is no federal funding for embryo research, but neither is there prohibition on privately funded research. This is because of the way we fund science. When the United States decided to allow states to regulate medical practice, rather than make medicine a matter of federal control, government leaders were acting on the grounds that local officials would better know what individual doctors were doing and thus be better able to oversee their practice. Though the U.S. government began medical research into diseases in 1887, at a small Staten Island laboratory of the Marine Services Hospital (largely inspired by growing awareness of infectious diseases and fears about immigration as a source of disease), federal control of medicine really only began in 1906.

After 1900, when most states had enacted some version of food quality oversight, a chemist for the Department of Agriculture, Dr. Harvey Wiley, lobbied for federal legislation to regulate food and drugs. His efforts brought about the Pure Food and Drug Act of 1906, which actually regulated relatively little: it ensured only that food and drugs contained what the manufacturer claimed and that it be made known if they contained narcotics. This legislation joined the Biologics Act of 1902, which had called for the federal Hygienic Lab (the lab that had begun on Staten Island and then moved to Washington), to exercise oversight of the production of vaccines and antitoxins. Though these laws established only small amounts of federal control of medicine, they represented the Progressive Era's move toward federal oversight and gave the government explicit authority to act in the medical interests of the American public.

The reasoning was that individual free-market choices could not work properly unless the consumer had sufficient information about the product. In some cases, information would require expertise and training, as well as access to laboratories, that most people did not have. Since the public health and even the life of citizens was at stake, federal regulation of medicine was warranted. In response to particular events, the food and drug regulatory legislation has been modified and expanded at several points since 1906, yet federal powers regarding medicine remain limited. The cabinet-level Department of Health, Education, and Welfare was established only in 1953, under President Eisenhower. Aside from matters of grave national importance, interstate commerce, and threats to national security, medicine remains a state interest.

While federal regulation has remained limited, however, federal funding for medical research has grown tremendously, from that minimal beginning on Staten Island to the most rapidly growing sector of the federal budget for research and development. Everybody loves medical research, it seems, even in the face of cuts and consolidations elsewhere. Even the fiscally conservative state of Arizona has managed to provide considerable financial incentive to attract a new Genomics Institute. Every state and every university feels the lure of biomedical research.

In 1930, the Ransdell Act, named for a Louisiana senator, brought important changes in federal attitudes to medicine. This legislation changed the name of the Hygienic Lab to the National Institute of Health, and it authorized federal funding for medical research and some basic biological research there. It brought about a considerable change in public attitudes toward medicine and increased recognition of the need to provide public support for research experts who could add to our body of knowledge about diseases and their treatments. The National Cancer Institute was authorized and funded in 1938, with the support of all senators in a rare show of unanimity. Despite the Depression and economic hardships, our nation's leaders felt it important to promote public health and the health of individual citizens. Research, they believed, would lead to new medical discoveries and therefore to important treatments good for all of us.

From this simple beginning, as a singular National Institute, Congress created more and more individual disease-oriented institutes. The organizational chart for today's multiple, coordinated National Institutes of Health requires an expert to untangle the reporting lines. With cries of "doubling the budget" for NIH ringing through the halls of Congress for the past several congressional sessions, the growth and complexity stand only to increase.

NIH funding is federal money, derived from tax dollars. The Congress appropriates such funds through dedicated appropriation bills for every fiscal year, and the institutes are charged to disperse them according to their respective missions and mandates. The general public and legislative assumption has been that research is good, discovery is good, and medical advance is good. While it has always been clear that we cannot publicly fund all research that might be useful, and even though some people cite concerns about distributive justice to argue that scarce public funds be more widely and equitably distributed, publicly funded research is nonetheless thought to be good.

Congressmen want us to know more about health, in order to make the lives of all Americans healthier and therefore presumably better.

To date, Congress has imposed few limitations on medical research expenditures, the only notable restriction being the use of federal funds for creation or destruction of human embryos. Researchers receiving federal support must follow all other laws, including guidelines for human and animal experimentation. While it has not funded all areas of science equally, of course, Congress has mandated only those few recent restrictions on funding. With a very few exceptions of abusive human experimentation, it has not prohibited research using private funds.

In 1945, Office of Science and Technology advisor Vannevar Bush presented President Truman with his report, published as *Science—The Endless Frontier*. In it he made the case for public support of scientific research through what was to become the National Science Foundation.[24] Creation of the NSF provided a powerful precedent for public funding of scientific research and for self-regulation of research by scientists through peer review. Understandably, scientists today do not want to see the principle of self-regulation violated through legislative or regulatory intervention from outside science. But some individuals and interest groups argue that it makes no sense to let the scientists be in control.

Previous prohibitions on federal funding for embryo research make up an important part of the background for these discussions. In 1974 with the National Research Act, Congress set in motion the steps that led to the National Commission for the Protection of Human Subjects five years later. That commission offered regulatory and oversight recommendations that led to the establishment of institutional review boards and an ethics advisory board. In 1979, the Ethics Advisory Board recommended that the United States develop a program of federally funded research using human embryos up to fourteen days old to assess the safety of *in vitro* fertilization. The board seems to have reasoned that we should carry out research trials rather than learn by trial and error on embryos implanted in actual patients. With similar reasoning, the United Kingdom developed the Human Fertilisation and Embryology Authority to oversee and administer embryo research.

In 1980, however, and before the research was ever developed in the United States, the Ethics Advisory Board's charter expired. This meant that while legislation was on the books to allow funding, there was no lawfully constituted body to review the protocols for the research. Without that review, funding could not proceed. In effect, then, the expiration of the charter meant the end of that federally funded research. Meanwhile, no systematic federal laws have governed fertility research and development in private clinics.

In 1988, the NIH decided to establish an internal panel, the Human Fetal Tissue Transplantation Research Panel (HFTTR). Maintaining that the ethics of abortion could be separated from arguments about using fetal tissue to treat diseases, this panel voted 18 to 3 that the NIH should fund embryo and fetal research. Yet the Secretary of the Department of Health and Human Services, which oversees the NIH, accepted the arguments of the three conservative members that research would increase the incentive for and therefore the number of abortions and extended what had become a moratorium on fetal and embryo research. President George H. W. Bush had agreed with the secretary and vetoed congressional efforts to end the moratorium and allow research to proceed with federal funding.[25] Some of the counterarguments in Congress focused on the preference that such research remain in the hands of the public rather than under the control of private firms.

The newly inaugurated President Clinton issued an executive order to lift the moratorium early in 1993. In 1994, an internal NIH panel supported research, but President Clinton then reversed his previous position in response to public pressure. Bioethicist Ronald Green, who served on the NIH panel, was clearly disgusted with Clinton's reversal. His account of the embryo wars throughout the 1990s provides an insider's look at the "vortex of controversy" and offers a thoughtful window on what is at issue on the side of allowing and publicly supporting research.[26]

The earlier National Bioethics Advisory Commission report had cautiously recommended federal funding for embryo research, including support for deriving embryonic stem cells and human embryonic germ cells. But the commission insisted on two major caveats and restrictions. Only two of the currently available sources of such cells should be allowed, namely already-dead fetal tissues and embryos (defined as fertilized eggs that have begun dividing and have the capacity to become humans) remaining after fertility treatments—precisely the sources that Gearhart and Thomson had, in fact, used.[27]

The commission also offered recommendations about how stem cells should be procured and what sorts of informed consent procedures would be best. Embryos and fetal tissue should never be bought or sold, for example. Finally, to make sure that all federally funded research conforms to the accepted ethical principles, there should be "an appropriate and open system of national oversight and review."[28] Other Washington groups, including the National Academy of Sciences and the American Association for the Advancement of Science, also held hearings, collected views, and produced reports, as did the NIH, and all reached similar recommendations. Since the legislation restricting the creation of or experimentation on embryos using federal funds had not prohibited research with embryos from private funds or donors, the NIH leadership accepted this interpretation and announced that they would begin accepting grant proposals for human embryonic stem cell research.

When the Department of Health and Human Services ruled in January 1999 that the NIH could allow such research to go forward, the public outcry was immediate and loud. Opposition, along with the fact that George W. Bush was campaigning for the presidency with declarations that he opposed human embryo research, including embryonic stem cell research, made researchers nervous. Few wanted to take up new projects that might well be stopped by political forces. Some researchers threatened to leave the United States for other countries, like the United Kingdom, where embryonic stem cell research was explicitly allowed. Others pointed to the opportunities that would be lost if the research was not carried forward and urged Congress for explicit funding support to proceed.

The President Decides about Stem Cells

Late in 2000, the NIH announced its final guidelines for federally funded embryonic stem cell research using stem cells derived with private funding. President Clinton issued an executive order in agreement. In the absence of congressional guidance, and given the considerable disagreements about even what was at issue, much less how the problems should be resolved, President Clinton concluded that using frozen embryos already destined to be discarded from fertility clinics was acceptable. As long as no federal funds were used to obtain or destroy the embryos or to derive the stem cell lines from the embryos, research on the cell lines themselves should meet the guidelines of existing legislation. Because Clinton issued his decision in an executive order, unless and until Congress overrode his decision, the order stood. Clinton left office and Bush arrived.

As the days of his first year in office ticked along, President Bush gave little indication of interest in scientific matters. He delayed filling the position of presidential science advisor until late, and then he demoted his choice as advisor and director of the Office of Science and Technology Policy, physicist John Marburger, to a lesser role than the one his counterparts had held in recent administrations. Other key science positions remained unfilled, yet the pressure continued to mount for Bush to make a decision about stem cell research. He had announced in his first month in office that the issue was under review.

By August 2001, after months of little pressing news and more than a month before the drastic events of September 11 riveted the nation's attention, Bush felt increasing pressure to reverse Clinton's decision about stem cell research. Fortunately, the NIH had recently issued a lengthy, thorough, thoughtful, and intelligent report on stem cells, *Stem Cells: Scientific Progress and Future Research Directions*.[29] That report is balanced and explains very clearly what the research involves, its potential benefits, its risks, and a range of issues involved. A *New York Times* story on August 11, 2001, included a picture of Bush advisor Karen Hughes holding the NIH report and saying that this was an important resource for the president's decision. President Bush decided to allow federal funding for research on stem cell lines that had already been derived by the time he spoke on August 9, 2001. No new lines could be created, even with private funds, and then studied with federal funding.

This might appear to be an act of hubris. It was as if the presi-

WHOSE VIEW OF LIFE?

dent were saying, "I have the power to decide and I decree that only stem cells developed prior to the moment of my decision right now can be used for research. Anything else cannot." He does not seem to have meant the decision that way, however. Nor was the August 10 *New York Times* editorial fair. Headlined "President Bush Waffles," it declared: "Last night George W. Bush had one of those rare opportunities a president gets to take a bold step that might define his administration. Instead he ducked." Rather, there is considerable evidence that the stem cell research issue was a source of internal conflict for Bush and that the decision reflected his attempt to lead. In the end, he satisfied no one. He did allow some research to go forward and left the door open for further support for more research. And he did not go so far as to crystallize immediate opposition in Congress. He had been placed in the political position where he felt he had to decide, and he decided.

In deciding, both Clinton and Bush were informed by numerous advisors. In his speech on August 9, Bush appointed his own bioethics council to advise him about stem cells and other bioscience matters. It had the mission of advising "the President on bioethical issues that may emerge as a consequence of advances in biomedical science and technology." Furthermore, "In connection with its advisory role, the mission of the Council includes the following functions: 1. to undertake fundamental inquiry into the human and moral significance of developments in biomedical and behavioral science and technology; 2. to explore specific ethical and policy questions related to these developments; 3. to provide a forum for a national discussion of bioethical issues; 4. to facilitate a greater understanding of bioethical issues; and 5. to explore possibilities for useful international collaboration on bioethical issues."[30]

Whereas Clinton's commission, as well as ethical advisory

committees at the NIH or the National Academy of Sciences, typically included a significant number of leading scientists, Bush's group consists more heavily of experts focused on the ethics, religion, or values implications of research rather than on the scientific details. This is not necessarily bad. But it is a decision about what counts as an expert.

Leon Kass (as chair) is particularly appealing to a conservative president, no doubt, because he holds an M.D. degree from the University of Chicago as well as a Ph.D. in biochemistry from Harvard and has very conservative social and moral views. His scientific credentials are impeccable, though he moved quickly from research in molecular biology to bioethics. As Kass put it, while looking back on the 1970s, "It dawned on me that there were large moral questions touching on human nature and human dignity that were being raised by these powerful and largely welcome developments in medical sciences." "It seemed to me that the real challenge of our society was to find a way to reap the benefits of new biology without sliding down the road to Brave New World and human degradation."[31]

Kass offers what he calls the "wisdom of repugnance" as a guide to what is morally good. If we find something repugnant (or if he does), then he concludes that it probably is unethical and undesirable. True, he realizes that "Revulsion is not an argument" and that furthermore, "some of yesterday's repugnances are today calmly accepted—not always for the better." Nonetheless, he says, "In some crucial cases, however, repugnance is the emotional expression of deep wisdom, beyond reason's power completely to articulate it. Can anyone really give an argument fully adequate to the horror that is father-daughter incest (even with consent), or bestiality, or the mutilation of a corpse, or the eating of human flesh, or the rape or murder of another human being?" "I suggest," Kass continues, "that our

WHOSE VIEW OF LIFE?

repugnance at human cloning belongs in this category." He realizes that a legislative ban on human cloning would be unprecedented. Nonetheless, he finds it not only fully warranted but a "golden opportunity" to avoid the "present danger," for "the humanity of the human future is now in our hands."[32]

One might accept Kass's policy conclusions in some cases but reject his problematic intuitionist approach, which relies on "what we find repugnant" and struggles to find legitimate grounding in a pluralistic society where people deeply disagree, and have good reasons for disagreeing, about what is repugnant. In fact, "repugnance" does not carry real wisdom, so we must import the "wisdom" and judgment from other sources. For the position of bioethicist-in-chief, President Bush selected Leon Kass not because of his perceived neutrality or wisdom for sorting through contested moral views. Rather, Kass was chosen precisely because he had a known view—and to Bush a politically acceptable view—on a contested issue. This approach may help a president decide his own position on a complex issue where science meets moral pluralism, and it may allow the president's view of life to prevail—for now. But it does not give us a way to negotiate wisely among the competing views and to provide effective ways to move toward stable bioscience policy.

As Clinton's ethics commission chairman, Harold Shapiro, reflected in an editorial essay: "Scientific progress is both planned and spontaneous, a science and an art, and is always subject to social, political, and cultural forces. Some of the influences on the scientific agenda originate within the science itself; others originate in the preferences, values, and aspirations of those who sponsor or finance scientific research." It was the responsibility of his commission to make recommendations about how to operate at this morally contested conjunction of scientific and social interests, and Shapiro took that role very seriously. Shapiro concluded that his commission's "deliberations (and those of professional societies, religious institutions, and town hall meetings) are part of an important and sustained public dialogue regarding the nature of the relationship between the evolving scientific agenda and important ethical considerations."[33]

Other groups also deliberated on the range of existing views. The National Research Council and Institute of Medicine carried out their own study and urged that "Studies with *human* stem cells are essential to make progress in the development of treatments for *human* disease, and this research should continue. Furthermore, stem cell research is just in its infancy and needs investment now, on both embryonic and adult stem cells. The work must be publicly-funded, with open peer review and debate about the science and its potential clinical applications." Finally, "In conjunction with research on stem cell biology and the development of potential stem cell therapies, research on approaches that prevent immune rejection of stem cells and stem cell–derived tissues should be actively pursued. These scientific efforts include the use of a number of techniques to manipulate the genetic makeup of stem cells, including somatic cell nuclear transfer."[34]

The Congress deliberated as well. On July 31, 2001, the United States House of Representatives passed a bill, H.R. 2505, the Human Cloning Prohibition Act of 2001. This legislation, which passed the House by a clear margin of 251 to 176, received little serious debate. The bill defines human cloning as "human asexual reproduction, accomplished by introducing nuclear material from one or more human somatic cells into a fertilized or unfertilized oocyte whose nuclear material has been removed or inactivated so as to produce a living organism (at any stage of development) that is genetically virtually identical to an existing or previously existing human organism." Somatic cell

WHOSE VIEW OF LIFE?

"means a diploid cell (having a complete set of chromosomes) obtained or derived from a living or deceased human body at any stage of development." With those definitions, the proposed law states that "It shall be unlawful for any person or entity, public or private, in or affecting interstate commerce, knowingly (1) to perform or attempt to perform human cloning; (2) to participate in an attempt to perform human cloning; or (3) to ship or receive for any purpose an embryo produced by human cloning or any product derived from such embryo."[35]

This bill was passed by the House in part to set the boundaries for decisions about stem cell research and to make it clear that even if such research were allowed, human cloning for the purposes of reproducing humans would be prohibited. The House rejected attempts to amend the bill in ways that would have limited its effect by keeping the ban on reproduction but allowing private companies to clone human embryos and carry out clinical research and development of therapies.

From the point of view of the House members, the language that remains prohibits cloning for any reasons, including therapeutic cloning. I suspect, however, that if this particular language were signed into law it would be open to many legal challenges. Some interpretations will hinge on the vagueness of our understanding of key terms. The precise meaning of the statement "so as to produce a living organism" is highly contested, for example. Indeed, it lies at the very heart of the debates about what is involved biologically and morally with cloning. If, in fact, nuclear transfer into an oocyte produces a hybrid that may be alive in some senses but is not a "life" in all senses, then perhaps the product is not really a "living organism" and not prohibited by this law. Much in the act remains to be interpreted, even if this legislation or something very much like it were passed into law.

As Congress-watchers often note, representatives in the House are occasionally quick to pass what they see as politically useful legislation that they know the Senate will not support. Thus, they are safe if they turn out to have been "wrong," but they are also safe if the Senate agrees and constituents cannot blame them alone. As the excellent *Washington Post* reporter Rick Weiss put it in a story the next day, opponents "characterized yesterday's action in the House as an easy opportunity for many lawmakers to prove their conservative credentials before going home to their constituents Friday for a month-long recess."[36]

Democratic Representative Peter Deutsch of Florida commented on his colleagues' preparation for the July 31 vote on cloning, including therapeutic cloning with somatic cell nuclear transfer, that this "may be the lowest level of knowledge I've seen . . . for a significant piece of legislation."[37] Deutsch expressed his own interpretation that a clone from a nuclear transplant "is not an embryo. It is not creating life by any definition of creating life."[38]

On February 27, 2003, the House of Representatives of the 108th Congress passed H.R. 534, the Human Cloning Prohibition Act, by a strong 241–155 vote and referred the bill to the Senate. By this time, the congressmen had had far more time to consider what they were doing and were far more informed about the implications of the legislation. This does not, of course, mean that they understood the underlying science any more deeply or that they were more open-minded about possible uses of cloning other than producing a full human being.

To date, the Senate has not enacted a parallel version of the bill and seems unlikely to do so. Action had been expected by May 2002, but was "indefinitely postponed by unanimous consent." The Senate bill that proponents of the ban favored would amend "the Public Health Service Act to prohibit: (1) perform-

　　　　　　WHOSE VIEW OF LIFE?

ing or attempting to perform human cloning; or (2) shipping, receiving, or importing the product of nuclear transplantation for the purpose of human cloning." Furthermore, it "provides that nothing in this Act shall be construed to restrict areas of biomedical, agricultural, and scientific research not specifically prohibited, including somatic cell nuclear transfer or other cloning technologies to clone molecules, DNA, cells, and tissues."[39] As so often happens with scientific issues, the bill gave way to issues considered to be more compelling national interests.

By 2003, the climate in the U.S. Senate had changed. Kansas Senator Sam Brownback introduced a bill like the one he had introduced the previous term, seeking to prohibit all human cloning. Yet Utah Senator Orrin Hatch had decided that the possible benefits of stem cell research outweigh the costs. He joined a coalition of senators including fellow Republican Pennsylvania Senator Arlen Specter and several Democrats, California Senators Barbara Boxer and Dianne Feinstein, Massachusetts Senator Edward Kennedy, Georgia Senator Zell Miller, Illinois Senator Richard Durbin, Iowa Senator Tom Harkin, and New Jersey Senator Frank Lautenberg, in proposing the Human Cloning Ban and Stem Cell Research Protection Act of 2003. This bill seeks to prohibit cloning for the purposes of creating a human being. But it explicitly allows research up to fourteen days in nonfertilized cloned eggs. As majority leader, Tennessee Republican Senator Bill Frist has indicated his interest in passing an anti-cloning bill, and he clearly favors a total ban. The discussions will clearly continue for some time whether Congress passes legislation or not, since the definitions and interpretations are subtle and it is not entirely clear who would have regulatory jurisdiction or oversight authority.

What is most important about these bills, more far-reaching than the particular legislation itself, is the boundary constraints that members of Congress want to establish. As eminent biologists Paul Berg, J. Michael Bishop, and Andrew Grove put it in an advertisement on May 5, 2002 (sponsored by the American Society for Cell Biology), "Congress Should Not Criminalize Medical Research." The trio invokes a powerful and undoubtedly misleading image: "Imagine a world where the cells in your body could be used to help cure you of cancer, diabetes, spinal cord injury, Alzheimer's and Parkinson's Diseases, and other medical problems. Now imagine a world where this miracle is denied to you, your parent or your child forever by Federal law." They go on to urge the Senate not to pass legislation outlawing stem cell research and nuclear transplantation, even if they do ban cloning humans for reproductive purposes. The ad was both compelling and a common strategy of political action. Both sides exaggerate and overemphasize their positions.

Political rhetoric is unfortunate, however, when it leads first-rate scientists known for their integrity and thoughtfulness to overstate their claims on behalf of the possible applicability of science. Of course we should do the scientific research, but research cannot guarantee that it will have the consequences we expect. Indeed, we might learn quite different things and we might be surprised in exciting new ways. The stakes here are extremely high. Prohibiting stem cell research would be as foolish as banning recombinant DNA experiments in the 1970s would have been. But suggesting that stem cell research using nuclear transfer is going to solve all those medical problems for all of us and our children gives the impression that scientific progress is inexorable.

Reporter Sheryl Gay Stolberg offers valuable insight when she notes that most stories pit the wondrous promise of science in bringing cures against the morality of religion. Yet what is partly at stake is a public fear of science, "and a corresponding desire to rein it in."[40] That seems exactly right. People love science and want all the greatest, latest advances that it may bring. Americans, at least in theory, also value free speech and freedom of inquiry, including through science. Yet they are also fearful of that which seems foreign or incomprehensible. The public may be loathe to let the scientists be the experts who oversee their own work, yet in recent years it has not been obvious that they trust the Congress to do a better job. Stolberg quotes Leon Kass as acknowledging that, "By and large, we Americans are enthusiasts for science" and apparently trust scientists. "But when science comes increasingly to work on the human body and mind, in ways that have the possibility of altering who we are, people are rightly concerned that their very humanity might be affected."

Competing Views of Life

Yet not everyone agrees that humanity is more likely to be affected in bad rather than good ways. That is why Kass's concept of "wisdom" based on repugnance cannot work. It is also why we must not foreclose research with the potential to help people who are also tax-paying American citizens. There are other interests at stake than those of the embryo and a minority of extremist pro-life lobbyists. The challenge is to develop ways of balancing the competing views and the competing interests. How can we decide whose interests are most important and how to make wise policy decisions for the future? While the supporters of therapeutic cloning or stem cell research lobbied through television ads featuring "Harry and Louise" chatting about the values of stem cell research, opponents featured their own "Harriet and Louis" worrying about the destruction of embryos. The viewer is left to wonder which of these neighborly couples is right.

In the face of competing views, we must not make policy out of reaction to something we do not understand or do not think we like even though we know little about it. We must not let a few voices outshout the many. We need a balance of enabling and protecting, of development and safety, of supporting the diversity of competing interests without allowing minority interests to prevail and win. We need open public discussion and richer shared understanding of what is involved in defining and refining life. We need deeper historical reflection on the context and conventions that we have come to accept. It will take good will, mutual respect for competing interests, and improved scientific literacy to act wisely.

This was the situation facing George W. Bush on August 9, 2001, when he acknowledged that deciding whether to fund stem cell research "forces us to confront fundamental issues about the beginnings of life and the ends of science." Furthermore, doing so leads us to dangerous ground where competing views contest what is right, for "as the discoveries of modern science create tremendous hope, they also lay vast ethical mine fields. As the genius of science extends the horizons of what we can do, we increasingly confront complex questions about what we should do. We have arrived at that brave new world that seemed so distant in 1932, when Aldous Huxley wrote about human beings created in test tubes in what he called a 'hatchery.'" "I'm a strong supporter of science and technology, and believe they have the potential for incredible good," he said, but "I also believe human life is a sacred gift from our Creator."

George W. Bush admitted that he was torn, and he sensed that the country was divided in its convictions. Seeking a compromise, he ended his speech "I have made this decision with great care, and I pray it is the right one." Bush allowed limited research to go forward, on only those few existing cell lines that completely arbitrarily existed at the moment he made his speech. The only possible argument in favor of that otherwise arbitrary line would be that there could be no incentive for researchers to rush to create new cell lines in order to meet some future deadline, though Bush did not actually make that point.

He also appointed a Council on Bioethics. This council would "keep us apprised of new developments and give our nation a forum to continue to discuss and evaluate these important issues," namely, "all of the medical and ethical ramifications of biomedical innovation."[41] In an opinion piece a few days later, on August 12, Bush continued: "Biomedical progress should be welcomed, promoted and funded—yet it can and must be humanized. Caution is demanded, because second thoughts will come too late. As we work to extend lives, we must do so in ways that preserve our humanity."[42]

Michael Gazzaniga, a member of that council, expressed evident dismay that Bush appeared to be making decisions before hearing what an expert panel had to say. Gazzaniga noted that when he "joined the panel, officially named the President's Council on Bioethics, I was confident that a sensible and sensitive policy might evolve from what was sure to be a cacophony of voices of scientists and philosophers representing a spectrum of opinions, beliefs and intellectual backgrounds. I only hope that in the end the president hears his council's full debate" and does not rush to judgment, as he appeared already to be doing.[43]

The President's Council on Bioethics Report

In July 2002, the council posted a draft of its report on its website and invited response.[44] The group offered the following recommendation as their "First Proposal":

We recommend a congressionally enacted ban on all attempts at cloning-to-produce-children and a four-year national moratorium (a temporary ban) on human cloning-for-biomedical-research. These measures would apply everywhere in the United States and would govern the conduct of all researchers, physicians, institutions, or companies, whether or not they accept public funding. We also recommend that, during this moratorium, the federal government undertake a thoroughgoing review of present and projected practices of human embryo research, pre-implantation genetic diagnosis, genetic modification of human embryos and gametes, and related matters, with a view to proposing, before the moratorium expires, an ethically acceptable public policy to govern these scientifically and medically promising but morally challenging activities. Several reasons converge to make this our recommended course of action at the present time. Members of the Council who support this recommendation do so for different reasons; some individual Members do not endorse all the concurring arguments given below.

The "Second Proposal" called, instead, for the following:

We recommend a congressionally enacted ban on all attempts at cloning-to-produce-children while preserving the freedom of cloning-for-biomedical-research. We recommend the establishment of a system of oversight and regulation that would permit cloning-for-biomedical-research to proceed promptly, but only under carefully prescribed limits. These measures would apply everywhere in the United States and would govern the conduct of all researchers, physicians, institutions, or companies, whether or not they accept public funding. In addition, we recommend that the federal government undertake a thorough-going review of present and projected practices of human embryo research. Several reasons converge to make this our recommended course of action at the present time. Members of the Council who support this recommendation do so for different reasons; some individual Members do not endorse all the concurring arguments given below.[45]

Ten members of the council voted in favor of the First Proposal, seven of whom had favored banning the research altogether plus three who favored a moratorium. Seven members voted in favor of the Second Proposal. This means there was a 7–3–7 split, in fact, with seven opposed and seven in favor of allowing cloning for purposes of allowing stem cell research. The vote demonstrates a significant split in this hand-picked group of experts, who had as many differences of opinion as there are among the educated public more widely.

Further reinforcing the point that the hand-picked council represents a range of divergent views, the initial predraft version of the report included an appendix consisting of twelve personal statements and more to come with the final version. These statements are strong and clear and express competing views of life. They make manifest how difficult it is to define life and to reconcile different views about the desirability of refining and improving life. And they reveal the challenges of negotiating an effective and wise set of conventional definitions and acceptable practices in a pluralistic society. It is hardly surprising that on August 9, 2001, George W. Bush admitted that consultation with many people revealed "widespread disagreement." He drew on prayer to guide his decision.

Conclusion

In a country founded on separation of church and state, it is not clear why it is prayer that should guide an American president to policy decisions about bioscience. In a country strengthened by the economic benefits of medical research and technology and its commitment to principles of freedom, the discordant calls of "people of many faiths" with different views about life should not outweigh the interests of those who might benefit from embryo research and its medical innovations without careful consideration and a high burden of reason for restricting inquiry.

Perhaps it will not be stem cell research that leads to major therapeutic advances but some other new knowledge that we discover only because this research was allowed and funded. Or we may decide that our health care and research dollars would be better spent improving preventative care and addressing chronic diseases. Perhaps we will decide that reproductive cloning carries too high a moral cost but that other embryo research is too potentially valuable not to fund. The point is that we do not know which of these policy decisions is wisest, nor whose view should prevail, because we have not had an informed and broad-ranging public discussion.

George W. Bush was exactly right in saying that second thoughts can come too late. That is true in many ways. We need a wiser approach to bioscience policy that is literate and informed both scientifically and socially. Stem cell research has gained enough public interest and understanding to allow at least open discussion of the Bioethics Council's Second Proposal: to prohibit reproductive cloning (cloning with the objective of producing an adult human being) but not to prohibit and even explicitly to allow therapeutic cloning (cloning with the objective of producing stem cells for research and clinical purposes). Actor Christopher Reeve's remarkable progress in regaining partial function after a severe spinal cord injury gives hope for what might be possible through regenerative medicine. Former first lady Nancy Reagan's quiet but clear support for stem cell research that might help Alzheimer's sufferers like her husband opens a door to public discussion for conservative Reagan supporters. Senator Orrin Hatch's new enthusiasm for stem cell research highlights what many see as the value in gaining knowledge in this area. Yet opinion polls and editorial commentary swing dramatically, depending on precisely how a question is worded and what the respondents think it means. It is not at all clear what "we the people" think or should think about embryo research.

As political analyst Adam Clymer noted, in a single week then-Congresswoman Connie Morella (a thoughtful and strong supporter of medical research and a moderate Republican whose Maryland district included the NIH) stated that Americans favored stem cell research three to one; the Juvenile Diabetes Foundation said it was 70 percent in favor; a Gallup poll gave 54 percent in favor; and the National Conference of Catholic Bishops found only 24 percent in favor.[1] Obviously, not only what question but also who is asked and who does the asking will make a difference to the results.

It will not suffice simply to put matters of science and medical research to a direct public vote. Unfortunately, the larger public remains generally nervous and uninformed about science, often preferring to let others make decisions about scientific matters. Despite educational efforts by the National Science Foundation, National Institutes of Health, and such public figures as Alan Alda or Bill Nye the Science Guy, all too high a percentage of the American public remains uninformed about science. As the NSF's *Science and Engineering Indicators* shows every other year when it is released, the public does not know much about science or about how science works. Instead, as astronomer Carl Sagan lamented, people seem actually to prefer to live in a "demon-haunted world" and to resort to superstition and psychics rather than to discover science.[2] We must change that by promoting public understanding of science, though this is just a first step toward wise biopolicy decision making.

Disagreements about how to develop intelligent medical and bioscience policy will obviously continue, and so they must. Such discussions can lead to wise and effective decisions if—but only if—they are first of all consistent with the best available understanding of science and grounded as well in respect for the existing diversity of views and a willingness to listen to others. Generally the most extreme views cannot lead to stability. The politics of hate and absolutism on any side will not move us wisely forward. Pitting a false picture of deeply entrenched but blind defenders of traditional morality and life against an equally false picture of amoral scientists ruthlessly pursuing their selfish ends is a bad way to make policy. False dichotomies only guarantee the political intractability of important issues on which some decision must be made.

So how are false dichotomies to be avoided? How are we to make the wisest medical and bioscience policy we can? On these issues, as we have seen in the preceding chapters, history has something to teach us, whether we are scientists, policy makers, citizens at large, or some combination of these. The first, and perhaps most crucial, thing we learn is that there have been many views of life and many ways to define when a life begins, just as there still are and always will be. The questions have changed as our science has advanced, and the answers have had to be renegotiated in the light of accumulating knowledge. Any policy answer should be grounded in the best available current science and the best moral thinking—and in the knowledge that science and morality are not intrinsically at odds. However much the current science may aid the deliberations, the answers have never followed from the scientific facts alone, nor from moral or religious intuitions alone.

The definitions are not obvious either, and have frequently been controversial, with competing views of life represented at the same time. These divergent definitions cannot each have been the one and only right way to view life. There is no reason to suppose that a "wisdom of repugnance" or other personal, intuitive pronouncements work any better as a moral guide now than they have in the past. In the long run we cannot just keep trying to outshout one another. The second lesson of history is that we need to learn humility and tolerance of one another's views. Decisions must be made, but that is precisely what they are: decisions or conventions of our own that can be reviewed and perhaps revised by each new generation and as new discoveries bring new questions.

Third, we have seen that we have felt ourselves on the brink of the abyss on several occasions provoked by scientific innovations. We have been convinced more than once that some new technology was just about to destroy life as we had known it—only to learn that the risks could be managed and that we were glad to have and use the new technology. If not every hope has panned out, neither has every fear.

We have also learned from past controversies a somewhat richer view of science. This is a fourth lesson from history: scientific views change in unexpected ways. The firm results of today become the discarded theories of tomorrow. This is not a drawback of science but its great strength—though it is also a reason for caution. Life-enhancing results may arise from unexpected pathways of research. Any science is more robust and progresses more rapidly when it is free to pursue different and competing avenues simultaneously. To foreclose any such avenue is a potential threat to our future well-being. Sometimes we must say no—but only for clear and compelling reasons.

Fifth, and finally, we have learned from history that in a democracy we must keep the discussion going among scientists, policy makers, and citizens at large. In a diverse society and a political system that reflects diversity, there may be no views of life that are universal, perhaps none that command even a majority. Yet some policy must be made, even if only for the time being. In such a situation we cannot foreclose the hopes for the many to forestall the fears of the few. This does not mean that scientists should have the license to pursue any path they wish while disregarding their fellow citizens. We must meet each other half-way. We must all learn more of the procedures and methods of science as well as of its promise and dangers. In short, we must all learn something of the language of science.

In addition, as the public must become more scientifically literate, at least some scientists must support this effort by learning to explain scientific processes and discoveries to a wider audience. Furthermore, more scientists must become politically literate; they must understand how policy is made and how moral disputes are adjudicated. They must learn to speak the language of policy and of democracy, to speak to our fears as well as our hopes, and to participate in the public and policy decisions about science.

There is room as well, in the broader academic community, for "interpreters" and "translators" of science. Communicating with a broad public audience is something that a so-called blue ribbon commission or hand-picked panel cannot always do well. There the members are selected precisely to be experts in their particular fields, with each chosen to represent a particular point of view. Likewise, the ethics and policy committees are set up to include a representative of every point of view deemed relevant by the person making the appointments, as the president's bioethics councils have been. Panelists chosen in this way may find it impossible to get beyond the clamor of differences or the conviction that they have been selected to hold certain views or play out certain roles.

In addition to these experts, we need wise interpreters who are capable of stepping back, looking at the bigger picture, and seeing how the pieces fit together. Training may be required to play, in effect, the role of negotiator or meta-expert, a more modest version of what used to be called a "Renaissance man." It is not obvious how we can get others to trust such translators or how to certify these meta-experts as such. But it would be worth trying. The judges might seek their help in deciding which sorts of experts to call on for a complex case; policy makers and congressional staff might consult with them as they sort through difficult and contested issues, in the way that the Office of Technology Assessment, for example, once advised Congress.

George W. Bush based his August 9, 2001, decision on his best available reports, scientific and moral knowledge, prayer, and instinct. In effect he had to play the role of negotiator himself. Yet he could not possibly be in a position to interpret all the views relevant to the decisions he was making; this is more than we can reasonably expect of any president. His advisors, a divided council, offered no clear guidance, but rather a range of divergent views and a split result to be interpreted. The history of events discussed in this book suggests that we need something different. We need to accept the pluralism of our democratic society but then to work with meta-experts to negotiate across divergent views of life. We need to forge a compromise explicitly responding to our best science and our best moral thinking, avoiding extremism and absolutism, and fully realizing that even our best current view of life should be expected to change and to keep changing.

Letting history teach us all these things is no panacea; it is certainly no guarantee that our controversies over medical and bioscience policy, over whose view of life is to guide us, will easily be resolved. Indeed, one can almost guarantee that the controversies will continue. Perhaps they should. But if we learn what we can, if we meet each other at a reflective point of mutual respect and tolerance, then surely our science will thrive and our policies will be wiser and more humane.

Notes

Introduction

1. See President Bush's speech at <www.whitehouse.gov/news/releases/ 2001/08/20010809–2.html>.
2. Legislation beginning with the FY1996 Balanced Budget Downpayment Acts 1 (P.L. 104–99) and each year since in appropriations bills covering the NIH.
3. Ceci Connolly, "Waging the Battle for Stem Cell Research," *Washington Post* (June 9, 2002).

1. From the Beginning

1. Aristotle, *Generation of Animals,* trans. A. L. Peck (Cambridge, Mass.: Harvard University Press, 1979), pp. 99, 111, 133, 129.
2. E. S. Russell, *The Interpretation of Development and Heredity: A Study in Biological Method* (1930; Freeport, N.Y.: Books for Libraries Press, 1972), pp. 11–26.
3. Aristotle, *Generation of Animals,* pp. 151–153, 157.
4. Laurie Zoloth, "The Ethics of the Eighth Day: Jewish Bioethics and Research on Human Embryonic Stem Cells," in Suzanne Holland,

Karen Lebacqz, and Laurie Zoloth, eds., *The Human Embryonic Stem Cell Debate* (Cambridge, Mass.: MIT Press, 2001), pp. 95–111.

5. St. Augustine, in his *Enchiridion*. This passage is widely cited, but I have not found the original wording. See "Abortion and Catholic Thought: The Little-Told History," a summary of Catholics for a Free Choice, *The History of Abortion in the Catholic Church*, reprinted in *Conscience* (Fall 1996) and available through several websites.

6. This history is cited in numerous places, as from P. Gasparri, ed., *Codicis iuris fontes* (Rome, 1927), p. 308.

7. C. D. O'Malley and J. B. de C. M. Saunders, *Leonardo da Vinci on the Human Body: The Anatomical, Physiological, and Embryological Drawings of Leonardo da Vinci* (New York: Greenwich House, 1982), pp. 470–485.

8. Monica Von During, Marta Poggesi, and Georges Didi-Huberman, *Encyclopedia Anatomica: A Complete Collection of Anatomical Waxes* (Los Angeles: Taschen Books, 1991).

9. William Harvey, *Exercitationes de Generatione Animalium* (London: William Dugard for Octavian Pullwyn, 1651).

10. David Bainbridge, *Making Babies: The Science of Pregnancy* (Cambridge, Mass.: Harvard University Press, 2000).

11. Richard S. Westfall, *The Construction of Modern Science* (Cambridge: Cambridge University Press, 1977), pp. 97–98.

12. *Encyclopaedia Britannica,* 1st ed., 1771.

13. Shirley Roe, *Matter, Life, and Generation* (Cambridge: Cambridge University Press, 1981); Peter Bowler, "Preformation and Pre-existence in the Seventeenth Century: A Brief Analysis," *Journal of the History of Biology* 4 (1971): 221–244; Peter Bowler, "The Changing Meaning of 'Evolution,'" *Journal of the History of Ideas* (1975) 36: 95–114.

14. See, e.g., Westfall, *Construction of Modern Science,* pp. 98–99; Roe, *Matter, Life, and Generation,* pp. 3–5.

15. See Roe, *Matter, Life, and Generation,* p. 176n1, for discussion of the distinction between "preformation" and "preexistence."

16. Malpighi, trans. Howard B. Adelmann, *Marcello Malpighi and the Evolution of Embryology,* 5 vols. (Ithaca, N.Y.: Cornell University Press, 1966), vol. 2, pp. 935–937.

17. Roe, *Matter, Life, and Generation,* p. 41.

18. Jane Maienschein, "Competing Epistemologies and Developmental Biology," in Richard Creath and Jane Maienschein, eds., *Biology and Epistemology* (Cambridge: Cambridge University Press, 2000), pp. 122–137, esp. pp. 123–126.

19. Karl Ernst von Baer, *De Ovi Mammalium et Homini Genesi* (1827). Thanks to Michael White for help with the translation.

20. Brian Bracegirdle, *A History of Microtechnique* (Ithaca, N.Y.: Cornell University Press, 1978); S. Bradbury, *The Microscope Past and Present* (Oxford: Pergamon Press, 1968).

21. Henry Harris, *The Birth of the Cell* (New Haven: Yale University Press, 1999), p. 120.

22. Karl Ernst von Baer, "Die Metamorphose des Eies Batrachier vor der Erscheinung des Embryo und Folgerungen aus ihr für die Theorie der Erzeugung," *Müller's Archiv für Anatomie, Physiologie und wissenschaftliche Medizin* (1834): 481–508.

23. Harris, *Birth of the Cell.*

24. Matthias Schleiden, *Beiträge zur Phytogenesis* (1838); Theodor Schwann, *Mikroskopsische Untersuchungen über die Übereinstimmung in der Struktur und dem Wachstum der Thiere und Pflanzen* (1839).

25. Schwann quoted in William Coleman, *Biology in the Nineteenth Century* (Cambridge: Cambridge University Press, 1977), p. 28. Also see Arthur Hughes, *A History of Cytology* (London: Abelard-Schuman, 1959).

26. Rudolf Virchow, *Cellularpathologie* (Berlin: August Hirschwald, 1858).

27. Cited in Harris, *Birth of the Cell,* p. 123.

28. George Newport, selected and arranged from the author's manuscript by George V. Ellis, "Researches on the Impregnation of the Ovum in

the Amphibia; and on the Early Stages of Development in the Embryo," *Philosophical Transactions of the Royal Society of London* (1854): 229–244.

29. Bracegirdle, *A History of Microtechnique,* provides a useful discussion of the development of microscopic techniques.

30. Charles Otis Whitman, *Methods of Research in Microscopical Anatomy and Embryology* (Boston: S. E. Cassino and Co., 1885).

31. See Fritz Baltzer, *Theodor Boveri: Life and Work of a Great Biologist,* trans. Dorothea Rudnick (Berkeley: University of California Press, 1967).

32. Oscar Hertwig, *Lehrbuch der Entwicklungsgeschichte der Menschen und der Wirbelthiere* (Jena: G. Fischer, 1898).

33. Edmund Beecher Wilson, *The Cell in Development and Inheritance* (New York: Macmillan, 1896; 2d ed. 1900; 3d ed. as *The Cell in Development and Heredity,* 1925); Wilson, *An Atlas of the Fertilization and Karyokinesis of the Ovum* (New York: Macmillan, 1895).

34. Jane Maienschein, "From Presentation to Representation in E. B. Wilson's *The Cell,*" *Biology and Philosophy* 6 (1991): 227–254.

35. Edmund Beecher Wilson, *The Physical Basis of Life* (New Haven: Yale University Press, 1923), pp. 4, 6, 45–47.

2. Interpreting Embryos, Understanding Life

1. Karl Ernst von Baer, *Über Entwickelungsgeschichte der Thiere,* trans. Thomas Henry Huxley as "On the Development of Animals, with Observations and Reflections" in Thomas S. Hall, ed., *A Source Book in Animal Biology* (Cambridge, Mass.: Harvard University Press, 1951), pp. 392–399, quotation p. 399. The meaning of this work is discussed by E. S. Russell in *The Interpretation of Development and Heredity* (1930; Freeport, N.Y.: Books for Libraries Press, 1972), pp. 34–38.

2. Ernst Haeckel, *Generelle Morphologie der Organismen* (Berlin: George Reimer, 1867).

3. There has been much debate about Haeckel's data over the past century, by evolutionary biologists and critics of evolution. That's the

problem with sloppy science: it can be used for unintended and illegitimate purposes. Michael Richardson and others drew attention to what they saw as Haeckel's "fraud" in 1997, and creationists immediately took up the charge and used it to undercut all evolutionary claims, as if these all rested in any sense on Haeckel's drawings from the nineteenth century! The science writer Elizabeth Pennisi summarized the discussions in "Haeckel's Embryos: Fraud Rediscovered," *Science* 277 (1997): 1435.

4. Charles Otis Whitman, *Methods of Research in Microscopical Anatomy and Embryology* (Boston: S. E. Cassino and Co., 1885).

5. Interview at the University of North Carolina with embryologist Donald Costello, 1967, placed in the Marine Biological Laboratory Archives, Woods Hole, Mass.

6. Jane Maienschein, "Cell Lineage, Ancestral Reminiscence, and the Biogenetic Law," *Journal of the History of Biology* 11 (1978): 129–158, discusses their reactions.

7. Oscar Hertwig, *The Biological Problem of To-Day: Preformation or Epigenesis? The Basis of a Theory of Organic Development,* trans. P. Chalmers Mitchell (1894; London: William Heinemann, 1896).

8. Wilhelm His, *Anatomie Menschlicher Embryronen* (Leipzig: F. C. W. Vogel): *I, Embryonen des Ersten Monaten* (1880); *II, Gestalt- und Grössenentwicklung bis zum Schluss des 2. Monats* (1882); *III, Zur Geschichte der Organe* (1885).

9. Nick Hopwood, "Producing Development: The Anatomy of Human Embryos and the Norms of Wilhelm His," *Bulletin of the History of Medicine* 74 (2000): 29–79, quotation p. 31. See also Hopwood's marvelous *Embryos in Wax: Models from the Ziegler Studio* (Cambridge: Whipple Museum of the History of Science, 2002).

10. Hopwood, "Producing Development," p. 39.

11. Franz Keibel and Franklin P. Mall, eds., *Manual of Human Embryology* (Philadelphia: Lippincott, 1910), p. xvi.

12. Ibid., p. xvii.

13. Ronan O'Rahilly and Fabiola Müller, *Developmental Stages in Human Embryos* (Washington: Carnegie Institution of Washington Pub-

lication 637, 1987). Adrianne Noe, Director of the National Museum of Health and Medicine, Armed Forces Institute of Pathology, Washington, D.C., has developed an extensive embryo archive and has written about the project, for example in Jane Maienschein, Marie Glitz, and Garland Allen, eds., *The Carnegie Institution of Washington's Department of Embryology* (Cambridge: Cambridge University Press, forthcoming). Also see Lynn M. Morgan, "Materializing the Fetal Body; or, What Are Those Corpses Doing in Biology's Basement?" in Lynn M. Morgan and Meredith W. Michael, eds., *Fetal Subjects, Feminist Positions* (Philadelphia: University of Pennsylvania Press, 1999), on the collection of embryos acquired in the 1920s through 1950s at Mount Holyoke College.

14. Keibel and Mall, *Manual,* pp. 980, xvi, 18.

15. Jane Maienschein, "The Origins of Entwicklungsmechanik," in Scott Gilbert, ed., *A Conceptual History of Modern Embryology* (New York: Plenum, 1991), pp. 43–61.

16. Wilhelm Roux, "Beiträge zur Entwickelungsmechanik des Embryo, No. 5. Über die künstliche Hervorbringung halber Embryonen durch Zerstörung einer der beiden ersten Furchungskugeln, sowie über die Nachentwickelun (Postgeneration) der fehlenden Körperhälfte," *Virchows Archiv für Pathologisches Anatomie und Physiologie und klinische Medizin* 114 (1888): 113–153. Translated in Benjamin Willier and Jane M. Oppenheimer, eds., *Foundations of Experimental Embryology* (New York: Hafner, 1964), pp. 2–37.

17. Hans Driesch, "Entwicklungsmechanische Studien. I. Der Werth der beiden ersten Furchungszellen in der Echinodermentwicklung. Experimentelle Erzeugen von Theil- und Doppelbildung," *Zeitschrift für wissenschaftliche Zoologie* 53 (1891–92): 160–178. Translated in Willier and Oppenheimer, *Foundations,* pp. 38–50.

18. Hans Spemann, *Embryonic Development and Induction* (New Haven: Yale University Press, 1938), esp. pp. 22–34.

19. August Weismann, *Das Keimplasm: Eine Theorie der Vererbung* (Jena: Gustav Fisher, 1892). Translated by W. Newton Parker and Harriet Rönnfeldt as *The Germ Plasm* (New York: Scribner's, 1893).

20. Thomas Hunt Morgan, "Regeneration in the Hydromedusa, *Gonionemus vertens*," *American Naturalist* 33 (1899): 939–951, quotation p. 951.

21. Thomas Hunt Morgan, "Chromosomes and Heredity," *American Naturalist* 64 (1910): 449–496.

22. Jacques Loeb, "On the Nature of the Process of Fertilization and the Artifical Production of Normal Larvae (Plutei) from the Unfertilized Eggs of the Sea Urchin," *American Journal of Physiology* 31 (1899): 135–138.

23. Jacques Loeb, "On the Nature of the Process of Fertilization," *Biological Lectures of the Marine Biological Laboratory at Woods Hole* (1899), in Donald Fleming, ed., *The Mechanistic Conception of Life* (Cambridge, Mass.: Harvard University Press, 1964), pp. 105–115, quotations pp. 114, 115.

24. Jacques Loeb, "The Mechanistic Conception of Life," *Popular Science Monthly* (1912), in Fleming, *Mechanistic Conception of Life,* pp. 5–34, quotations pp. 8, 16, 28, 32.

25. Philip Pauly, *Controlling Life: Jacques Loeb and the Engineering Ideal in Biology* (New York: Oxford University Press, 1987).

26. Harrison provides a valuable survey of this topic in "Heteroplastic Grafting in Embryology," presented initially as a lecture on Dec. 14, 1933, and published in *The Harvey Lectures* (Baltimore: Williams and Wilkins, 1935), pp. 116–157. Reprinted as chap. 6 in Ross Granville Harrison, *Organization and Development of the Embryo,* ed. Sally Wilens (New Haven: Yale University Press, 1969).

27. Donna Jeanne Haraway, *Crystals, Fabrics, and Fields: Metaphors of Organicism in Twentieth-Century Developmental Biology* (New Haven: Yale University Press, 1976).

28. Hans Spemann and Hilde Mangold, "Über die Induktion von Embryonalanlagen durch Implantation artfremder Organisatoren," *Archiv für Entwicklungsmechanik der Organismen* 100 (1924): 599–638. On Mangold's role, see Viktor Hamburger, *The Heritage of Experimental Embryology: Hans Spemann and the Organizer* (New York: Oxford University Press, 1988), pp. 173–180.

29. Hamburger, *Heritage of Experimental Embryology.*
30. Ross G. Harrison, "The Outgrowth of the Nerve Fiber as a Mode of Protoplasmic Outgrowth," *Journal of Experimental Zoology* 9 (1910): 787–846.

3. Genetics, Embryology, and Cloning Frogs

1. See Fritz Baltzer, *Theodor Boveri: Life and Work of a Great Biologist,* trans. Dorothea Rudnick (Berkeley: University of California Press, 1967). Also, on Boveri in light of later developmental biology, see Jane M. Oppenheimer, "Questions Posed by Classical Descriptive and Experimental Embryology," in John A. Moore, ed., *Ideas in Modern Biology,* vol. 6 of *Proceedings of the XVI International Congress of Zoology* (Garden City, N.Y.: Natural History Press, 1965), pp. 206–227.
2. R. A. Fischer, "Has Mendel's Work Been Rediscovered?" *Annals of Science* 1 (1936): 115–137.
3. Gregor Mendel, "Versuche über Pflanzen-Hybriden," *Verhandlungen des naturforschenden Vereines in Brünn* 4 (1865): 3–47, appeared in 1866. For a reprint and interpretation, see Curt Stern and Eva R. Sherwood, eds., *The Origin of Genetics: A Mendel Source Book* (San Francisco: Freeman, 1966).
4. For anyone interested in Mendel, the MendelWeb internet site is very useful. See <www.netspace.org/MendelWeb/homepage.html>.
5. Henry Harris, *The Birth of the Cell* (New Haven: Yale University Press, 1999), p. 172.
6. Thomas Hunt Morgan, "Recent Theories in Regard to the Determination of Sex," *Popular Science Monthly* 64 (1903): 97–116, quotation p. 116. Also see Jane Maienschein, "What Determines Sex? A Study of Converging Approaches, 1880–1916," *Isis* 75 (1984): 457–480.
7. Thomas Hunt Morgan, "What Are 'Factors' in Mendelian Explanations?" *American Breeders' Association Reports* 5 (1909): 365–368, quotation p. 366.
8. Ibid., p. 365.

9. Thomas Hunt Morgan, "Chromosomes and Heredity," *American Naturalist* 44 (1910): 449–496, quotation p. 451.

10. Robert E. Kohler, *Lords of the Fly: Drosophila Genetics and the Experimental Life* (Chicago: University of Chicago Press, 1994).

11. On this work, see esp. Garland E. Allen, *Thomas Hunt Morgan* (Princeton: Princeton University Press, 1978); and Kohler, *Lords of the Fly.*

12. Horace Freeland Judson, *The Eighth Day of Creation* (New York: Simon and Schuster, 1979); Robert Olby, *The Path to the Double Helix: The Discovery of DNA* (Seattle: University of Washington Press, 1974).

13. J. D. Watson and F. H. C. Crick, "A Structure for Deoxyribose Nucleic Acid," *Nature* 171 (1953): 737.

14. George Bernard Shaw, *Man and Superman: A Comedy and a Philosophy* (Cambridge, Mass.: The University Press, 1903), paragraph 8 of "The Revolutionist's Handbook and Pocket Companion."

15. Diane B. Paul, *Controlling Human Heredity: 1865 to the Present* (Atlantic Highlands, N.J.: Humanities Press, 1995).

16. Daniel Kevles, *In the Name of Eugenics: Genetics and the Uses of Human Heredity* (Cambridge, Mass.: Harvard University Press, 1995); Philip R. Reilly, *A History of Involuntary Sterilization in the United States* (Baltimore: Johns Hopkins University Press, 1991).

17. Mark H. Haller, *Eugenics: Hereditarian Attitudes in American Thought* (New Brunswick: Rutgers University Press, 1963 and 1984), p. 141.

18. Stephen Jay Gould, "Carrie Buck's Daughter," *Natural History* 93 (1984): 14–18, p. 17.

19. *Buck v. Bell,* 247 U.S. 300 (1927).

20. William Carlos Williams, "A Night in June," in *The Farmers' Daughters: Collected Stories* (Norfolk, Conn.: New Directions, 1961).

21. Thomas Hunt Morgan, *Experimental Embryology* (New York: Columbia University Press, 1927), p. 8.

22. Thomas Hunt Morgan, *Embryology and Genetics* (New York: Columbia University Press, 1934), p. 17.

23. Robert E. Kohler, *Partners in Science: Foundations and Natural Scientists, 1900–1945* (Chicago: University of Chicago Press, 1991).
24. Hans Spemann, *Embryonic Development and Induction* (New Haven: Yale University Press, 1938), p. 211.
25. Ibid., p. 369.
26. Ibid., p. 367.
27. Robert Briggs, E. U. Green, and T. J. King, "An Investigation of the Capacity for Cleavage and Differentiation in *Rana pipiens* Eggs Lacking 'Functional' Chromosomes," *Journal of Experimental Zoology* 116 (1951): 455–499.
28. Robert Briggs and T. J. King, "Transplantation of Living Nuclei from Blastula Cells into Enucleated Frog's Eggs," *Proceedings of the National Academy of Sciences* 38 (1952): 455–463, quotations p. 455.
29. Ibid., p. 463.
30. Herbert John Webber, "New Horticultural and Agricultural Terms," *Science* 18 (1903): 501–503.
31. Briggs and King, "Transplantation," p. 456.
32. J. B. Gurdon and Alan Colman, "The Future of Cloning," *Nature* 402 (1999): 743–746, for an overview. See also Anne McLaren, "Cloning: Pathways to a Pluripotent Future," *Science* 288 (2000): 1775–1780.
33. James Watson, John Tooze, and David Jurtz, *Recombinant DNA: A Short Course* (New York: Scientific American Books, 1983), pp. 207–208.

4. Recombinant DNA, IVF, and Abortion Politics

1. "Recombinant DNA: A Critic Questions the Right to Free Inquiry," *Science* 194 (1976): 303–306.
2. Sheldon Krimsky, *Genetic Alchemy: The Social History of the Recombinant DNA Controversy* (Cambridge, Mass.: MIT Press, 1982).
3. Gordon Research Conferences at <www.grc.uri.edu>.
4. June Goodfield, *Playing God: Genetic Engineering and the Manipulation of Life* (New York: Random House, 1977), p. 93.
5. The NIH has had a series of changes to its Recombinant DNA Advi-

sory Committee (RAC), which has been updated periodically and reauthorized since 1975.

6. On this issue, see esp. Nicholas Wade, *Ultimate Experiment: Man-Made Evolution* (New York: Walker, 1977).

7. Vannevar Bush, *Science: The Endless Frontier* (Washington: Government Printing Office, 1945), p. 10.

8. Samuel Florman, *The Existential Pleasures of Engineering,* 2d ed. (New York: St. Martin's, 1996).

9. See Ira H. Carmen, *Cloning and the Constitution: An Inquiry into Governmental Policymaking and Genetic Experimentation* (Madison: University of Wisconsin Press, 1985), chap. 3.

10. Maxine Singer, "The Recombinant DNA Debate," *Science* 196 (1977): 127.

11. Harlyn Halvorson, "Recombinant DNA Legislation: What Next?" *Science,* no. 198 (1977): 357.

12. Philip H. Abelson, "Recombinant DNA Legislation," *Science* 199 (1978): 135.

13. James Watson, John Tooze, and David Jurtz, *Recombinant DNA: A Short Course* (New York: Scientific American Books, 1983), preface.

14. Charles Weiner, "Drawing the Line in Genetic Engineering: Self-regulation and Public Participation," *Perspectives in Biology and Medicine* 44 (2001): 208–220, quotation p. 218; citing Leon Kass, "Babies by Means of *in vitro* Fertilization: Unethical Experiments on the Unborn?" *New England Journal of Medicine* 285 (1971): 1174–1179.

15. Carl Feldbaum, "Some History Should Be Repeated," *Science* 295 (2002): 975.

16. Adele Clarke, *Disciplining Reproduction: Modernity, American Life Sciences, and "the Problems of Sex"* (Berkeley: University of California Press, 1998), pp. 81, 122.

17. Nicolas Rasmussen, "Biotechnology before the 'Biotech Revolution,'" in Carsten Reinhart, ed., *Chemical Sciences in the Twentieth Century* (Weinheim: Wiley-VCH, 2001), pp. 201–227.

18. Miles R. McCarry, "Doing What Comes Artificially," *Invention & Technology* (Summer 1999): 34–41.

19. Cited in Merriley Borell, "Biologists and the Promotion of Birth Control Research, 1918–1938," *Journal of the History of Biology* 20 (1987): 51–87, p. 60.

20. Gregory Pincus, *The Control of Fertility* (New York: Academic Press, 1965), p. 3.

21. See Miles Weatherall, *In Search of a Cure* (Oxford: Oxford University Press, 1990), chap. 5.

22. See Peter Singer and Deane Wells, *Making Babies: The New Science and Ethics of Contraception* (New York: Scribner's, 1985); originally published in 1984 in the United Kingdom by Oxford University Press as *The Reproduction Revolution: New Ways of Making Babies.*

23. Singer and Wells, *Making Babies,* p. vii.

24. Robert Edwards and Patrick Steptoe, *A Matter of Life: The Story of a Medical Breakthrough* (New York: Morrow, 1980), chap. 22.

25. Ibid., p. 155.

26. Ibid., pp. 179, 180.

27. Rebecca Mead, "Eggs for Sale," *New Yorker* (Aug. 9, 1999): 56–65, quotation p. 58.

28. Lori B. Andrews, "The Sperminator," *New York Times Magazine* (Mar. 28, 1999): 62–65.

29. Cited in Carey Goldberg, "Massachusetts Case Is Latest to Ask Court to Decide Fate of Frozen Embryos," *New York Times* (Nov. 5, 1999): A19.

30. Singer and Wells, *Making Babies,* p. viii.

31. David Bainbridge, *Making Babies: The Science of Pregnancy* (Cambridge, Mass.: Harvard University Press, 2001).

32. Leon Kass, cited in Singer and Wells, *Making Babies,* pp. 46–47, from *HEW Support of Research Involving Human In Vitro Fertilization and Embryo Transfer,* Appendix, section 2, quotations from pp. 21, 32.

33. Gina Kolata, "Fertility, Inc.: Clinics Race to Lure Clients," *New York Times* (Jan. 1, 2002), late edition, F1.

34. See Nicola Beisel, *Imperiled Innocents: Anthony Comstock and Family Reproduction in Victorian America* (Princeton: Princeton University Press, 1998).

35. Elizabeth Siegel Watkins, *On the Pill: A Social History of Oral Contraceptives, 1950–1970* (Baltimore: Johns Hopkins University Press, 1998), p. 29.

36. Etienne-Emile Baulieu with Mort Rosenblum, *The "Abortion Pill" RU-486: A Woman's Choice* (New York: Simon and Schuster, 1991), p. 15.

37. *Roe v. Wade,* 410 U.S. 113 (1973), IX B.

38. Baulieu, *"Abortion Pill,"* pp. 15, 16.

39. Gail Collins, "Celebrating One Hundred Years of Failure to Reproduce on Demand," *New York Times* (Apr. 14, 2002), sec. 4: 12.

40. J. F. Fletcher, "Knowledge, Risk and the Right to Reproduce: A Limiting Principle," in Aubrey Milunsky and George J. Annas, eds., *National Symposium on Genetics and the Law II* (New York: Plenum, 1980), pp. 131–135; quotation p. 131. Thanks to Diane Paul for bringing this to my attention in her excellent unpublished essay, "From Reproductive Responsibility to Reproductive Autonomy."

41. Bentley Glass, "'Science': Endless Horizons or Golden Age?" *Science,* no. 171 (1971): 23–29, quotation p. 28.

42. Michelle Andrews, "Birth Control Is Changing and Its Price Is Falling," *New York Times* (Apr. 21, 2002), late edition, Final, sec. 3, 8.

5. From Genetics to Genomania

1. President William Jefferson Clinton, "Remarks by the President, The Entire Human Genome Project," Office of Science and Technology Policy, Washington, June 6, 2000. Cited in Nicholas Wade, *Lifescript: How the Human Genome Discoveries Will Transform Medicine and Enhance Your Health* (New York: Simon and Schuster, 2001), as indicating the project's significance, and widely cited in print and televised media that day.

2. Lewis Carroll, *Alice's Adventures in Wonderland & Through the Looking-Glass* (New York: Heritage Press, 1941), p. 35.

3. Daniel J. Kevles, "Out of Eugenics: The Historical Politics of the Human Genome," in Daniel J. Kevles and Leroy Hood, eds., *The Code of Codes: Scientific and Social Issues in the Human Genome Project*

(Cambridge, Mass.: Harvard University Press, 1992), pp. 3–36, quotation p. 15.

4. See Anne Fausto-Sterling, *Sexing the Body: Gender Politics and the Construction of Sexuality* (New York: Basic Books, 1999).

5. See Horace Freeland Judson, *The Eighth Day of Creation* (New York: Simon and Schuster, 1979); also Judson's updated summary of advances, "A History of the Science and Technology behind Gene Mapping and Sequencing," in Kevles and Hood, eds., *Code of Codes,* pp. 37–80. James D. Watson, *The Double Helix: A Personal Account of the Discovery of the Structure of DNA* (New York: Atheneum, 1968). For a different perspective, see Brenda Maddox, *Rosalyn Franklin: Dark Lady of DNA* (New York: HarperCollins, 2002).

6. *Oxford English Dictionary,* online ed., "Genomics" <www.oed.com>.

7. John Beatty, "Origins of the U.S. Human Genome Project: Changing Relationships between Genetics and National Security," in Phillip R. Sloan, ed., *Controlling Our Destinies: Historical, Philosophical, Ethical, and Theological Perspectives on the Human Genome Project* (Notre Dame, Ind.: University of Notre Dame Press, 2000), pp. 131–153. For the most complete and thoughtful account of the origins of the genome project, see Robert Cook-Deegan, *The Gene Wars: Science, Politics, and the Human Genome* (New York: Norton, 1994).

8. Renato Dulbecco, "A Turning Point in Cancer Research: Sequencing the Human Genome," *Science* 231 (1986): 1055–1056.

9. Leslie Roberts, "Controversial from the Start," *Science* 291 (2001): 1182–1188.

10. Roger Lewin, "Proposal to Sequence the Human Genome Stirs Debate," *Science* 232 (1986): 1598–1600.

11. Ibid., p. 1598.

12. Ibid., p. 1600.

13. Bruce Alberts, "Introduction," in National Research Council, *Mapping and Sequencing the Human Genome* (Washington: National Academy Press, 1988), p. viii.

14. *Mapping and Sequencing the Human Genome,* p. 85.
15. Ibid., p. 103.
16. Roberts, "Controversial from the Start." p. 1184.
17. Michael Waldholz and Hilary Stout, "Rights to Life: A New Debate Rages over the Patenting of Gene Discoveries," *Wall Street Journal* (Apr. 17, 1992), A1.
18. John Burris, Robert Cook-Deegan, and Bruce Alberts, "The Human Genome Project after a Decade: Policy Issues," *Nature Genetics* 20 (Dec. 20, 1998): 333–335.
19. David Magnus, Arthur Caplan, and Glenn McGee, eds., *Who Owns Life?* (Amherst, Mass.: Prometheus, 2002).
20. Nancy Wexler, "Clairvoyance and Caution: Repercussions from the Human Genome Project," in Kevles and Hood, eds., *Code of Codes,* pp. 211–243, quotation p. 243.
21. Ruth Hubbard, "Genomania and Health," *American Scientist* 83 (1995): 8–10.
22. John Opitz, "Afterword: The Geneticization of Western Civilization: Blessing or Bane?" in Sloan, ed., *Controlling Our Destinies,* pp. 429–450.
23. James D. Watson, "A Personal View of the Project," in Kevles and Hood, eds., *Code of Codes,* pp. 164–173, quotation p. 173.
24. Eliot Marshall and Elizabeth Pennisi, "Hubris and the Human Genome," *Science* 280 (1998): 994.
25. For an excellent account of the genome race, see Kevin Davies, *Cracking the Genome: Inside the Race to Unlock Human DNA* (New York: Free Press, 2001).
26. Gerald Weissman, *The Year of the Genome: A Diary of the Biological Revolution* (New York: Henry Holt, 2002), p. 68.
27. Lee M. Silver, *Remaking Eden: How Genetic Engineering and Cloning Will Transform the American Family* (New York: Avon Books, 1997).
28. Quoted in Wade, *LifeScript,* p. 176.
29. Ibid., p. 178.
30. See, e.g., Francis Fukuyama, *Our Posthuman Future: Consequences*

of the Biotechnology Revolution (New York: Farrar, Straus and Giroux, 2002); James C. Peterson, *Genetic Turning Points: The Ethics of Human Genetic Intervention* (Grand Rapids, Mich.: Eerdmans, 2001); Brian Tokar, ed., *Redesigning Life: The Worldwide Challenge to Genetic Engineering* (London: Zed Books, 2001); Jon Turney, *Frankenstein's Footsteps: Science, Genetics, and Popular Culture* (New Haven: Yale University Press, 1998). Reviews of the various questions include Carl F. Cranor, *Are Genes Us? The Social Consequences of the New Genetics* (New Brunswick: Rutgers University Press, 1994); Lori Andrews, *Future Perfect: Confronting Decisions about Genetics* (New York: Columbia University Press, 2001); and Glenn McGee, *The Perfect Baby: A Pragmatic Approach to Genetics* (Lanham, Md.: Rowman and Littlefield, 1997).

31. "Whole-istic Biology," *Science* 295 (2002): 1661–1682. "It's Not Just in the Genes," *Science* 296 (2002): 685–703.

6. Facts and Fantasies of Cloning

1. Gina Kolata, *Clone: The Road to Dolly and the Path Ahead* (New York: William Morrow, 1998), p. 29.

2. Roslin Institute, Edinburgh, 1996/1997 Annual Report, accessible on the Institute website <www.roslin.ac.uk/publications/9697annrep/dollymania2.html> and in releases from 1997.

3. Lee M. Silver, *Remaking Eden: How Genetic Engineering and Cloning Will Transform the American Family* (New York: Avon Books, 1997). Gina Kolata, "Scientist Reports First Cloning Ever of Adult Mammal," *New York Times* (Feb. 23, 1997), final edition, 1.

4. Michael Specter and Gina Kolata, "After Decades of Missteps, How Cloning Succeeded," *New York Times* (Mar. 3, 1997), final ed., A1.

5. For discussion of these discoveries, see I. Wilmut, K. Campbell, and C. Tudge, *The Second Creation: The Age of Biological Control by the Scientists Who Cloned Dolly* (London: Headline Press, 2000).

6. Thomas H. Murray, "Even If It Worked, Cloning Wouldn't Bring Her Back," *Washington Post* (Apr. 8, 2001), B1.

7. Lee Silver, "What Are Clones?" *Nature* 412 (2001): 21. On definitions of cloning, see also Jane Maienschein, "What's in a Name: Embryos, Clones, and Stem Cells," *American Journal of Bioethics* 2 (2002): 12–19; and Ursula Mittwoch, "'Clone': The History of a Euphonious Scientific Term," *Medical History* 46 (2002): 381–402.

8. See, e.g., Richard Lewontin, "Cloning and the Fallacy of Biological Determinism," in Barbara MacKinnon, ed., *Human Cloning: Science, Ethics, and Public Policy* (Urbana: University of Illinois Press, 2000), pp. 36–48, quotation p. 48.

9. Robert A. Weinberg, "Of Clones and Clowns," *Atlantic Monthly* (June 2002): 54–59.

10. "Name-calling Gets Stem-Cell Researcher into Hot Water," *Nature* 419 (2002): 4.

11. Richard Dawkins, "What's Wrong with Cloning?" in Martha C. Nussbaum and Cass R. Sunstein, eds., *Clones and Clones: Facts and Fantasies about Human Cloning* (New York: Norton, 1998), pp. 54–66, quotation p. 54.

12. The Nuremberg Code, in *Trials of War Criminals before the Nuremberg Military Tribunals under Control Council Law no. 10,* vol. 2 (Washington: Government Printing Office, 1949), pp. 181–182. Available at <http://ohsr.od.nih.gov/nuremberg.php3>.

13. The National Commission for the Protection of Human Subjects of Biomedical and Behavioral Research, "The Belmont Report: Ethical Principles and Guidelines for the Protection of Human Subjects of Research," Apr. 18, 1979. Available at <http://ohsr.od.nih.gov/mpa/belmont.php3>.

14. Kolata, *Clone,* p. 34.

15. Leon Kass, "The Wisdom of Repugnance: Why We Should Ban the Cloning of Humans," *New Republic,* June 2, 1997, pp. 17–26.

16. Glenn McGee and Arthur Caplan, "What's in the Dish?" *Hastings Center Report* 29 (1999): 36–38.

17. Roger Gosden, *Designing Babies: The Brave New World of Reproductive Technology* (New York: Freeman, 1999).

18. Rudolf Jaenisch and Ian Wilmut, "Don't Clone Humans," *Science* 291 (2001): 2552.

19. Worries about "telomere" length in cloned sheep continue to emerge, for example, despite the lack of clear evidence about whether there is such a phenomenon or whether it matters to development. We just do not know enough to be sure.

20. Ronald M. Green, *The Human Embryo Research Debates: Bioethics in the Vortex of Controversy* (New York: Oxford University Press, 2001).

21. Philip Kitcher, "There Will Never Be Another You," in MacKinnon, ed., *Human Cloning*, pp. 53–67, quotation p. 67; rpt. from Kitcher, *The Lives to Come* (New York: Simon and Schuster, 1997). See also Kitcher, *Science, Truth, and Democracy* (New York: Oxford University Press, 2001).

22. See, e.g., Nussbaum and Sunstein, eds., *Clones and Clones,* part 4. Lori Andrew has explored legal and ethical questions in a number of books, including *Body Bazaar: The Market for Human Tissue in the Biotechnology Age,* with Dorothy Nelkin (New York: Crown, 2001), and *The Clone Age: Adventures in the New World of Reproductive Technology* (New York: Henry Holt, 1999).

23. President Bill Clinton to Dr. Harold Shapiro, Feb. 24, 1997, published with the report "Cloning Human Beings: Report and Recommendations of the National Bioethics Advisory Commission," Rockville, Md., June 1997.

24. Kyla Dunn, "Cloning Trevor," *Atlantic Monthly* (June 2002): 31–52.

25. Gerald Weissman, *The Year of the Genome: A Diary of the Biological Revolution* (New York: Henry Holt, 2002).

26. National Academy of Sciences, National Academy of Engineering, Institute of Medicine, and National Research Council, *Scientific and Medical Aspects of Human Reproductive Cloning* (Washington: National Academy Press, 2002), p. 1. For many of the key papers and discussion of cloning, see Michael Ruse and Aryne Sheppard, eds.,

Cloning: Responsible Science or Technomadness? (New York: Prometheus, 2001).

27. See Andrea L. Bonnicksen, *Crafting a Cloning Policy: From Dolly to Stem Cells* (Washington: Georgetown University Press, 2002).

7. Hopes and Hypes for Stem Cells

1. For the National Marrow Donor Program, see <www.marrow.org/NMDP/history_stem_cell_transplants.html>.

2. Edmund Beecher Wilson, *The Cell in Development and Inheritance* (New York: Macmillan, 1896), p. 111.

3. R. L. Gardner, "Mouse Chimaeras Obtained by Injection of Cells into the Blastocyst," *Nature* 220 (1968): 596–597.

4. R. G. Edwards, "IVF and the History of Stem Cells," *Nature* 413 (2000): 349–351.

5. On the role of cytoplasm, see Jan Sapp, *Beyond the Gene: Cytoplasmic Inheritance and the Struggle for Authority in Genetics* (New York: Oxford University Press, 1987).

6. Helen Pearson, "Dual Identities," *Nature* 417 (2002): 10–11. Figure 8 in Chapter 7, by artist Margaret C. Nelson, was drawn after "The Body's Master Builders," an illustration from the *New York Times,* Science Times, Tues., Dec. 18, 2001, special section on stem cells; the sources for the illustration were Dr. Ronald D. G. McKay and Dr. Nadya Lumelsky, National Institutes of Health, and Dr. Nora Sarvetnick, Scripps Research Institute.

7. Bert Vogelstein, Bruce Alberts, and Kenneth Shine, letter to the editor, *Science* 297 (2002): 52.

8. Lewis Wolpert, *Triumph of the Embryo* (New York: Oxford University Press, 1991), p. 12. Michael Gazzaniga cited in "Metaphor of the Week," *Science* 295 (2002): 1637.

9. James Thomson et al., "Embryonic Stem Cell Lines Derived from Human Blastocysts," *Science* 282 (1998): 1145–1147.

10. M. J. Shamblott et al., "Derivation of Pluripotential Stem Cells from Cultured Human Primordial Germ Cells," *Proceedings of the National Academy of Sciences* 95 (1998): 13726–13731; J. D. Gearhart,

"New Potential for Human Embryonic Germ Cells," *Science* 282 (1998): 1061–1062.

11. Gretchen Vogel, "Can Old Cells Learn New Tricks?" *Science* 287 (2000): 1418–1419. See Stuart H. Orkin and Sean J. Morrison, "Stem-Cell Competition," *Nature* 418 (2002): 25.

12. Constance Holden and Gretchen Vogel, "Plasticity: Time for a Reappraisal?" *Science* 296 (2002): 2126–2129.

13. Jose B. Cibelli et al., "Parthenogenetic Stem Cells in Nonhuman Primates," *Science* 295 (2002): 819.

14. Kyla Dunn, "Cloning Trevor," *Atlantic Monthly* (June 2002): 31–52, quotation p. 49.

15. Paul Berg, "Progress with Stem Cells: Stuck or Unstuck?" *Science* 293 (2001): 1953; Irving L. Weissman and David Baltimore, "Disappearing Stem Cells, Disappearing Science," *Science* 292 (2001): 601.

16. Irving L. Weissman, "Stem Cells: Scientific, Medical, and Political Issues," *New England Journal of Medicine* 346 (2002): 1576–1579.

17. Arthur Caplan, "Half a Loaf Is Not Good Enough," *Scientist* (Sept. 17, 2001): 6.

18. Arthur L. Caplan, "Attack of the Anti-Cloners" [Comment], *Nation* (June 17, 2002).

19. See discussions of ongoing research, such as the special section "Stem Cells Branch Out," *Science* 287 (2000): 1417–1446, or the Howard Hughes Medical Institute's special issue of its *Bulletin* (March 2002), entitled "Stem Cells to the Rescue." For a more popular treatment, see Stephen S. Hall, "The Recycled Generation," *New York Times Sunday Magazine* (Jan. 30, 2000).

20. Kathleen Burge, "Science of Evidence Puts Judges to the Test," *Boston Sunday Globe* (May 19, 2002), quoting Superior Court Judge Ralph Gants.

21. For an irreverent look at the role of experts, see Sheldon Rampton and John Stauber, *Trust Us, We're Experts* (New York: Putnam, 2002).

22. On the legacy of the OTA, see <www.wws.princeton.edu/~ota/>.

23. For a thoughtful look at the possible reasons, see Rick Weiss, "Free to Be Me: Would-Be Cloners Pushing the Debate," *Washington Post* (May 12, 2002): A1.

24. Vannevar Bush, *Science: The Endless Frontier* (Washington: Government Printing Office, 1945).

25. For a succinct summary, see Dorothy C. Wertz, "Embryo and Stem Cell Research in the USA: A Political History," *Trends in Molecular Medicine* 8 (2002): 143–146.

26. Ronald M. Green, *The Human Embryo Research Debates: Bioethics in the Vortex of Controversy* (New York: Oxford University Press, 2001).

27. National Bioethics Advisory Commission, *Ethical Issues in Human Stem Cell Research* (Rockville, Md., Sept. 1999). Available now through the Georgetown website at <http://bioethics.georgetown .edu/nbac/> since the commission's mandate expired Oct. 1, 1999.

28. Ibid.

29. National Institutes of Health, *Stem Cells: Scientific Progress and Future Research Directions: Opportunities and Challenges: A Focus on Future Stem Cell Applications* (June 2001). See <www.nih.gov/news/ stemcell/scireport.htm>.

30. See the mandate for President Bush's Council on Bioethics at <www.bioethics.gov>.

31. Pam Belluck, "The President's Decision: The Bioethicist; Trying to Balance Science and Humanity," *New York Times* (Aug. 11, 2001), final edition, A11.

32. Leon Kass, "Why We Should Ban Human Cloning Now: Preventing a Brave New World," *New Republic* (May 2001), available at <www.tnr.com/052101/kass052101>.

33. Harold T. Shapiro, "Ethical Dilemmas and Stem Cell Research," *Science,* no. 285 (1999): 2065.

34. National Research Council and Institute of Medicine, *Stem Cells and the Future of Regenerative Medicine* (Washington: National Academy Press, 2001). Also see the related report, National Academy of Sciences, National Academy of Engineering, Institute of Medicine, and

National Research Council, *Scientific and Medical Aspects of Human Reproductive Cloning* (Washington: National Academy Press, 2002).

35. See the *Congressional Record,* for legislative bills and other supporting documents, at <http://thomas.loc.gov>.

36. Rick Weiss, "House Votes Broad Ban on Cloning," *Washington Post* (Aug. 1, 2001), A1.

37. Representative Deutsch quoted in *Science* 293 (Aug. 10, 2001): 1043.

38. Gerald Weissman, *The Year of the Genome: A Diary of the Biological Revolution* (New York: Henry Holt, 2002), p. 17.

39. *Congressional Record,* various bills were read and referred to committee on Health, Education, Labor, and Pensions in March 2002. See <http://thomas.loc.gov>.

40. Sheryl Gay Stolberg, "It's Alive! It's Alive!" *New York Times* (May 5, 2002), final edition, sec. 4, 16.

41. President George W. Bush, "Remarks on Stem Cell Research," Aug. 9, 2001. Available at <www.whitehouse.gov/news/releases/2001/08/20010809-2.html>.

42. George W. Bush, "Stem Cell Science and the Preservation of Life," *New York Times* (Aug. 12, 2001), late edition, sec. 4, 13.

43. Michael S. Gazzaniga, "Zygotes and People Aren't Quite the Same," *New York Times* (Apr. 25, 2002), late edition, A31.

44. See <www.bioethics.gov> for the council's report and related information.

45. President's Council on Bioethics, "Human Cloning and Human Dignity: An Ethical Inquiry," July 2002, at <www.bioethics.gov>.

Conclusion

1. Adam Clymer, "Wrong Number: The Unbearable Lightness of Public Opinion Polls," *New York Times* (July 22, 2001).

2. Carl Sagan, *The Demon-Haunted World: Science as a Candle in the Dark* (New York: Random House, 1996).

Index

Aristotle *(continued)*
    gradualist view of life's beginning, 13–16, 25, 26, 33; on mixing of male and female fluids, 34, 49; on process of formation, 87; theory of the soul, 14–16
Arrowsmith (Lewis), 103
Artificial insemination (AI), 140, 141
Artificial reproductive technology (ART), 153
Autonomy issues, 164, 224, 234

Baltimore, David, 131, 270
Barry, Martin, 39–40
Baulieu, Etienne-Emile, 164–165
Behr, Barry, 154
Belmont Report, 231
Berg, Paul, 131, 269–270, 292
Bioethics, 11, 12, 153, 166, 200, 223, 229, 230–240, 247, 285, 286; expertise, 200–201. *See also* President's Council on Bioethics
Biologics Act, 277–278
Biomedical research and technology, 18, 139, 278
Bioscience policy, 12, 194, 287–288, 298, 300–304
Biotechnology, 138, 139, 140, 172
Birth control, 140, 141, 143, 155–160, 164, 166–168, 238. *See also* Contraception
Birth Control Clinical Research Bureau, 156
Bishop, J. Michael, 292
Blackmun, Justice, 162
Blair, Tony, 170
Blastocyst, 146, 248, 259, 260, 261
Blastopore, 70, 71, 76, 84, 118
Blood, 173, 252; transfusions, 251

Bone marrow transplants, 251, 252–253
Bonnet, Charles, 29
Born, Gustav, 68, 80–81
Botstein, David, 183, 187
Boveri, Theodor, 44, 89, 90, 94, 172
Brenner, Sydney, 187
Briggs, Robert, 118–119, 120–122, 123, 212, 221; nuclear transfer experiments, 127, 219
Brown, Louise Joy, and family, 144, 146, 147–148, 150, 151, 230, 245, 254, 269
Brownback, Sam, 6, 291
Buck, Carrie and Doris, 104–107
Bush, George H. W., 191, 192, 281
Bush, George W., 1–2, 6, 9, 152, 167, 271; decision about stem cell research, 283, 284–285, 286, 287, 294–295, 297, 298, 304; response to cloning, 242
Bush, Vannevar, 133, 280

Caltech, 110, 125, 173, 179
Cancer, 182, 251, 292
Caplan, Arthur, 232–233, 235, 271–272
Capron, Alexander, 151
Carnegie Institution (Washington), 65, 86, 115
Carrel, Alexis, 86
Carruthers, Marvin, 179
Catholic Church, 158, 159, 164; interpretation of when life begins, 17–18, 31, 42, 261
Celera, 171, 201–202, 203, 205, 206
Cell(s): continuity from other cells, 39, 40; cultures, 255, 261, 265;

cytology, 39, 41, 43–47; differentiation, 40, 41, 43, 44, 119, 221, 257, 260, 261–262; division, 38, 39–40, 45, 46, 49, 57, 59, 69–70, 72, 76, 80, 89, 90, 115, 220, 256–257, 259; fixing and preserving, 42–43; as fundamental unit of life, 39, 40, 48; genetic material in, 31, 48, 113; membranes, 37, 69; movement/migration, 81, 82, 257–259; potencies in, 116. *See also* Stem cell(s); Stem cell research

Cell lines, 57, 59, 60, 247, 255, 260, 263, 265–266, 270–271, 294–295

Cell nuclei, 38, 39–40, 43, 46, 47, 76, 89, 118, 119, 120, 122, 123, 256

Cell theory, 35, 36–48

Chromosomes, 44–46, 73, 88–90, 111, 256; genetic material in, 45, 89, 93, 94, 100, 111, 173–174; male, or Y, 206; sex determination by, 100, 173

Civil rights issues, 108, 161

Clinton, Bill, 192, 217; cloning issues and, 240–241, 242, 247; decision about stem cell research, 276, 283–287; on Human Genome Project, 170–171; reversal of position on human embryo research, 282

Cloning, 3, 71, 115–118, 120, 176, 246–248, 267; ban on federal funding for, 246–247; benefits and risks, 232, 236, 240, 246; of endangered species, 243–244; evolution and, 129, 236, 244; of frog eggs, 118–119, 212; of genes, 128, 241; as genetic replication, not exact copy, 120, 227–228, 229, 234,

235, 249; mistaken conceptions about, 222–226; policy and politics, 139, 223, 238, 240–242, 247; public funding for, 237; reproductive, 139, 222, 238, 269; role of the media, 224; rush to clone after Dolly, 242–244; social implications, 216, 225, 247, 248. *See also* Dolly the sheep; Ethics of cloning; Human embryo research; Human reproductive cloning; Therapeutic cloning

Cloning Rights United Front, 238

Cold Spring Harbor Laboratory, 183, 191, 192

Collins, Francis, 169, 170, 171, 203, 205, 208; appearance before congressional committee, 201, 202–204, 205

Collins, Gail, 165–166

Comstock laws, 155–156, 161

Conception, 79, 140; as beginning of the individual life, 235; epigenetic, 96; as moment of life, 3, 7–9, 25, 42, 257. *See also* Preformationism

Contraception, 5, 140–141, 155–157, 158, 160, 162, 164; moral issues surrounding, 18, 160. *See also* Birth control

Cook-Deegan, Robert, 194, 195

Council on Bioethics, 233, 295–297, 299

Court rulings: abortion, 164, 238; birth control, 156–157, 238; egg and sperm ownership, 150–151; scientific testimony in court, 274–276; sterilization, 104–107; when life begins, 162, 163. *See also specific cases*

Crick, Francis, 101, 191

Edwards, Robert, 145, 146, 147, 148, 151, 155, 254

Egg(s), 16, 24, 32, 57, 72, 76, 88, 95, 165, 256; as cell, 40, 41, 44, 49, 93; as continuity with the mother, 32, 33, 41, 44; development, 34–35, 37, 143; donation of, 149, 268; enucleated, 115, 116, 121–122, 219, 288; fertilization by sperm 57, 78; fixing and preserving, 34–35; freezing of, 149, 151; frog, 35, 81–82; harvesting of, 145–146, 148; inherited from mother, 49, 87, 172; legal ownership and custody debates, 148, 149–151; lines of cells generated by, 265–266; mammalian studies, 32–36; market for and sale of human, 149, 151, 239, 268; mouse, 145, 146; nuclear transfer, 219–220, 221, 237, 246, 288, 289; nuclear transplantation, 115–124; release from ovary, 141; sex hormones and, 143; theory that form resides in, 27–28. *See also* Fertilization

Ehlers, Vernon, 241, 242, 249

Eisenhower, Dwight D., 133, 278

Embryo, 67, 118; change to fetus, 50, 60; cloned and implanted, 139; collections of, 65–66; definition of, 4, 9, 24–25, 60, 260; "designer," 149, 152; ensoulment of, 17, 18; formed at fertilization, 29; freezing of, 149, 272; implantation in the uterus, 4, 9, 65; laws governing, 151; pre-embryos, 30; preimplantation, 30, 260; presence of beating heart in, 29–30; religious interpretation of, 16–19; restrictions on creation or destruc-

tion of, 280, 283; transfer, in cows, 140–141, 142; transfer to surrogate sheep mother, 220

Embryogenesis, 36, 49, 57, 59, 88; representational images of, 60–67

Embryology, 13, 41, 57, 67, 87, 88, 110–115, 123, 145; Darwin on, 52–53; definition of terms, 260–261; early attempts to see inside the body, 19–32; early theories of egg and sperm, 27–29; experimental, 83, 85, 87, 113, 299; explantation, 72, 85–87; human, 61–64; transplantation and, 118

Embryonic development, 56, 175; beginning of the individual life, 54, 55–56, 57, 138; blastula stage, 59, 76, 123, 142, 164, 260; causes of, 77–78, 83; environmental conditions and, 56, 66, 69–70, 75, 79–80, 81, 83, 95–96; formation of major body parts and organs, 65, 66–67; gastrula stage, 56, 57, 59, 70, 76, 119, 260, 261; germ layers, 56, 57, 59; Mendelian-chromosomal theory of inheritance and, 95–97, 113, 127; morula stage, 59, 260; normal, 50, 60, 64, 66; parallels in structures, 52–53, 54–55, 56, 58; patterns, 51, 57, 58, 60, 70–73, 83; as process of formation, 53, 57, 60; role of external condition, 76–80; stages of, 4, 51, 53, 59–65, 66, 75, 143, 255–262; standardization in, 60, 64

Ensoulment. *See* When life begins (debate): ensoulment theory (and quickening)

Epigenesis, 8, 15, 73, 82, 95, 112, 256, 257

Ethics of cloning, 216, 225, 230, 232, 234, 248, 289; public concern and debate over, 153, 228, 237–239, 243, 245, 248. *See also* Bioethics; *specific areas of debate*

Eugenics, 101–110, 112, 159, 165–166; Human Genome Project and, 198

European Molecular Biology Laboratory, 184

Evolution, 33, 52–54, 223; adaptation and, 56; cloning issues and, 129, 236; conservative nature of, 209; education, 229; hybridization and, 127

Expertise, 136, 144, 229, 233, 278; admissibility of scientific testimony in court, 274–276; bioethical, 200–201; across fields, 223; medical, 109; in policy making, 196; for presidential decisions about stem cell research, 286

Feinstein, Dianne, 291

Fertility clinics, 5, 153, 235, 254, 256, 263, 268, 281, 282

Fertilization, 3, 9, 44, 60, 78, 93, 139–140, 256, 266; artificial, 115; as beginning of life, 42, 235; as mixing of male and female fluids, 43, 44; outside the mother, 5, 10, 146, 148, 150, 151, 256. *See also* Egg(s); *In vitro* fertilization (IVF); Sperm

Fetus, 262; definition of, 9, 25, 50, 60; formation of major body parts in, 65, 66–67, 262; Fourteenth Amendment protection of, 162,

163; viability of (life outside the mother), 9, 163

Flemming, Walther, 45

Fletcher, Joseph, 166

Fol, Hermann, 43, 45

Food and Drug Administration (FDA), 158, 242

Frist, Bill, 291

Galton, Francis, 102

Gazzaniga, Michael, 261, 262, 295

Gearhart, John, 263–264, 282

Gene(s), 59, 173, 256; as carriers of heredity, 112–113; in cells, 113; chromosomes and, 111; cloning of, 128, 209, 241; defects and mutations, 112, 173–174, 207, 256; definition of, 177–178, 206; definition of life and, 224, 225

Genentech, 138

Generation, 13, 14, 23–24, 44, 49; spontaneous, 14, 27, 38, 39, 40

Genetic determinism, 31, 223–226, 256

Genetic engineering, 220, 243, 246

Genetics, 52, 82, 87, 101, 111, 112, 113, 145, 172, 177, 178, 256; embryology and, 110–115; ethical and legal issues surrounding, 238; human, 173

Genome, 89, 173, 179. *See also* Human Genome Project

Genomics, 172, 178, 179, 206, 237, 239; rush to, 198, 204

Geron Corporation, 263

Gilbert, Walter, 182–183

Glass, Bentley, 166

Goodell, Margaret, 264–265

Human Genome Project *(continued)* of DOE in, 180, 183, 184, 186, 190; role of NIH in, 171, 177, 180, 185, 186, 190, 191, 202, 203; social implications, 189, 207–208; technology transfer to industry and business, 194, 203

Human reproductive cloning, 247–248, 272, 276, 298; ban on federal funding for, 241, 242; benefits of, 243, 250; claims of, 244–245; definition of, 288; legislative ban on, 242, 247, 249, 287, 289, 292, 298; to produce replacement tissues or cells, 249–250, 271–272; public concern over, 217–218, 219, 222–223, 224–225, 236, 241. *See also* Cloning; Dolly the sheep; Therapeutic cloning

Hybrid organisms, 81, 82, 92, 212, 257; from egg and sperm of different donors, 148

Hygienic Act, 277–278

Implantation in the uterus, 4, 9, 141–142, 262; of blastocysts, 260; of cloned cells, 248; of cloned embryo, 139; preimplantation, 30; prevention of, 157, 158

Individual rights, 108, 109, 164

Induction, 83–85, 84, 117

Infertility, 144–145, 149, 152, 153, 161, 268; cloning and, 242

Inheritance, 47, 71, 69, 101; dominant and recessive factors, 91–94; law of independent assortment, 93, 94; law of segregation, 93; Mendelian-chromosomal theory of, 89,

90–93, 95–101, 113, 172; role of chromosomes in, 94; through genes, 127. *See also* Heredity

Institute to Sequence the Human Genome, 181

Internal institutional review boards, 231–232

Intuitionism, 233–234, 287, 301

In vitro fertilization (IVF), 5, 145–154, 165, 256, 260, 269; ethical issues surrounding, 230; lack of regulation of, 153, 154; laws governing, 149, 150; public opinion of, 151–152; research, 281

Islamic interpretation of when life begins, 18

Jacobson v. Massachusetts, 106

Jaenisch, Rudolf, 236

James S. McDonnell Foundation, 186

Jewish law and theology, 16–17, 31, 261

Juvenile Diabetes Foundation, 299

Kass, Leon, 139, 153, 233, 293; wisdom of repugnance, 152, 286–287, 293, 301

Keibel, Franz, 64–66

Keller, Evelyn Fox, 210

Kennedy, Edward, 126, 135, 136, 186, 291

Kevles, Daniel, 103, 173

King, Thomas J., 118–119, 120–122, 123, 127, 212, 221; nuclear transfer experiments, 123, 219

Kitcher, Philip, 223, 237–238

Kohler, Robert, 99

Kolata, Gina, 153–154, 214–215, 216, 217, 218
Kurtz, David, 123, 137

Lander, Eric, 185
Lanza, Robert, 243
Laparoscopy, 145–146
Lawrence Livermore Laboratory, 184
Leonardo da Vinci, 19
Leukemia, 251, 252
Lewin, Roger, 183
Lewontin, Richard, 223, 227, 234
Lillie, Frank, 140
Loeb, Jacques, 76–79, 115, 265, 273
Los Alamos National Laboratory, 182, 184, 186

Mall, Franklin Paine, 65, 66
Malpighi, Marcello, 27–29
Mangold, Hilde, 84
Manhattan Project, 186
Marine Biological Laboratory at Woods Hole, Mass., 57, 76, 77
Massachusetts Institute of Technology (MIT), 128, 129
Materialism/mechanism, 26–27, 32, 39, 53, 78–79
McCarry, Miles, 140–141
McCormick, Katherine Dexter, 158
McGee, Glenn, 235
McKusick, Victor A., 179, 187
Mead, Rebecca, 149
Medical advances, 2, 12, 108, 164, 204, 251, 270, 279, 286

Mendel, Gregor, 47, 89, 90–93, 94, 98, 101, 172
Mendelian-chromosomal theory of inheritance, 89, 90–93, 95–101
Microscope(s), 21, 27, 29, 34, 35, 37–38, 41–43; used in embryology, 58, 64
Microtome cutting tool, 42, 61, 64
Mohr, Otto, 141
Molecular biology, 127, 177, 178, 179, 181, 183, 191, 223, 224; Human Genome Project and, 225
Morella, Connie, 299
Morgan, Thomas Hunt, 73, 74–75, 76, 94–95, 118, 123, 178, 273; on embryology, 96, 110–111; on embryology and genetics, 112–114; fruit fly studies, 97–100; gene-chromosome correlation, 172; on Mendelian-chromosomal theory of inheritance, 95–101; Nobel Prize awarded to, 112; studies of sex determination, 95–96
Morphogenesis/morphology, 49, 51, 52, 60, 67, 78
Multipotency, 254, 255
Munson, Lewis, 216
Murray, Thomas, 225
Muslim interpretation of when life begins, 31, 261
Mutation: in chromosomes, 206; in fruit flies, 97–100; in genes, 112, 173–174, 207, 256; sex-specific, 99–100

Nathans, Daniel, 127, 187
National Academy of Sciences, 126, 131, 135, 218, 229, 247, 276,

National Academy of Sciences *(continued)* 282–283; ethical advisory committees, 286; panel on stem cell research, 270

National Association of Science Writers, 169, 218

National Bioethics Advisory Commission (NBAC), 240, 242, 246–247, 276, 282

National Cancer Institute, 279

National Center for Human Genome Research, 190, 191

National Commission for the Protection of Human Subjects of Biomedical and Behavioral Research, 231, 280

National Conference of Catholic Bishops, 164, 300

National Family Planning and Reproductive Health Association, 161–162

National Institute of General Medical Sciences, 190

National Institutes of Health (NIH), 2, 115, 131, 132, 205, 208, 278–279, 300; ethical advisory committees, 285–286; funding for research, 137, 265, 279, 283; guidelines for recombinant DNA procedures, 135, 136–137; Human Fetal Tissue Transplantation Research Panel, 281; patents for genome sequences, 194–195; report on stem cells, 264–265; research strategies, 180; role in Human Genome Project, 171, 177, 180, 185, 186, 190, 191, 202, 203

National Laboratory Gene Library Project, 184

National Marrow Donor Program, 252

National Museum of Health and Medicine, 65–66

National Organ Transplant Act, 253

National Research Act, 231, 280

National Research Council, 115, 187, 218, 287

National Right to Life Committee, 164

National Science Foundation (NSF), 185, 280, 300

Nature, 169, 170, 172, 205, 215, 222; articles on Dolly the sheep, 213, 214

Nazi policies, 107, 230

Neel, James V., 172–173, 180–181

Newport, George, 41–42

New York Times: coverage of Dolly the sheep, 214, 217, 218; reports on stem cell research, 266, 284, 285

Nixon, Richard M., 159, 167

Nuclear transplantation, 115–124, 119–120, 122, 212, 219, 237, 246, 289, 290, 292

Nye, Bill, 300

Observer (newspaper), 214, 215

Office of Science and Technology Policy, 276

Office of Technology Assessment (OTA), 275–276, 303

Opitz, John, 198

O'Rahilly, Ronan, 66

Oregon Regional Primate Research Center, 243

Organisms, 223; complexity of, 210; definition of, 67, 117; human, 246;

Ruddle, Francis H., 179, 187
Rusconi, Mauro, 34

Sachs, Julius, 76
Sagan, Carl, 300
Salmon, Matt, 241, 249–250
Sanger, Margaret, 156, 158
Schleiden, Matthias, 36, 37, 38, 39, 43–44
Schwann, Theodor, 36, 37, 38, 39, 43–44
Science: education, 134, 229; funding for, 110, 171; lack of public understanding of, 136, 300, 302–303; place in society, 2–3, 5, 133, 134, 237; policy, 237, 240–242, 245, 250, 294; and the public, 218, 226–229, 292–293; role of the media in reporting, 213–219; "sound science" concept, 228–229
Science, 125, 131, 136, 169, 170, 172, 182, 205, 210, 263, 270; stem cell research published in, 251, 264
Science Service, 218, 226
Scripps, E. W., 218
Searle Pharmaceutical Company, 158
Sedgwick, William Thompson, 47, 253
Sex determination, 95–96, 100, 140
Shapiro, Harold, 240, 287–288
Shaw, George Bernard, 102
Sickle-cell anemia, 173
Silver, Lee, 207, 215–216, 226, 228
Singer, Maxine, 130, 134, 135–136, 183
Singer, Pete, 143, 144, 151
Sinsheimer, Robert, 125–127, 133, 134, 181
Sixtus V, Pope, 18

Slosson, Edwin E., 140, 218
Smith, Hamilton O., 127
Smithsonian Institution, 231
Society for Developmental Biology, 123
Somatic cells, 45, 123, 221, 237, 243, 244, 246, 248, 267, 288–290
Soul: Aristotelian theory of, 14–18; presence of, 7, 14–15. *See also* When life begins (debate): ensoulment theory (and quickening)
Specter, Arlen, 291
Specter, Michael, 217
Spemann, Hans, 72, 80, 81, 83, 84, 87, 121; early cloning research, 115–118
Sperm: as cell, 45; extenders, 140; inherited from father, 49, 87, 172; joining with egg, 5, 34, 41, 57, 78; legal ownership and custody debates, 149; theory of homunculus in, 27. *See also* Fertilization
Spock, Benjamin, 108
Stem cell(s), 38, 75; adult, 264–265, 288; clinical treatments using, 260; cultured human embryonic, 263; differentiation, 252, 263; embryonic, 254, 260, 264, 265, 276–277; hematopoietic, 252; importance to organ transplantation, 253; lines, 255; neural, 266; pluripotent, 266; specialized cells from, 253–254; therapeutic potential of, 265
Stem cell research, 71, 239, 262–264, 294; benefits and risks/costs, 277, 284, 289, 291; Bush's decision on, 284–285, 286, 287, 294–295, 297, 298, 304; combined with cloning,

250, 266; debates over, 153, 272; definition of life and, 3; early cloning research and, 116; early history, 251–255; federal funding for, 1–2, 6, 7, 265, 277; funding, 276–283; legislation governing, 238, 264, 271, 298; National Academy of Sciences panel on, 270; private funding for, 265, 283; prohibition on, 292; pro-life movement opposition to, 267, 269; public concern over, 228, 276, 299–300; public funding for, 262, 277, 288; report of the President's Council on Bioethics, 295–297, 299; sources of cells used in, 282; using human cells, 288; using mice, 253–254, 266

Steptoe, Patrick, 145, 146, 147, 148, 151, 155, 254

Stolberg, Sheryl Gay, 292

Strauss, Michael, 169, 170

Sutton, Walter, 172

Therapeutic cloning, 6, 250, 266–274, 294, 299; federal funding for, 268; legislation, 290; opposition to, 267, 269, 270, 271; prohibition of, 289; public approval of, 276–277; public funding for, 268–269

Thomas Aquinas, Saint, 17–18

Thomson, James, 263, 282

Tilghman, Shirley, 187, 189

Tooze, John, 123, 137, 187

Totipotency, 71, 117, 118, 254, 257

Transplantation, 72, 80–85, 88, 115; in early cloning research, 116–117; embryology and, 118; of fertilized eggs, 146; of nuclei, 115–124; as precursor to cloning, 120; rejection of foreign cells, 266, 267

Truman, Harry S, 280

Twins, 234–235, 241

United Kingdom, 248, 281, 283

University of California at Santa Cruz (UCSC), 181

University of Chicago, 87

University of Michigan, 135

Uterus, 143, 146. *See also* Implantation in the uterus

Van Leeuwenhoek, Anton, 27, 34, 41

Variation, 47, 91, 97–100

Varmus, Harold, 191, 202, 234–235

Ventner, Craig, 169, 170, 171, 180, 201–203, 205

Vesalius, Andreas, 19, 22

Virchow, Rudolf, 39

"Visible Embryo Project," 66

Vitalism, 24, 25–26, 29, 30, 31, 38, 48

Vogelstein, Bert, 261

Von Baer, Karl Ernst, 32, 33, 34–35, 36, 39, 51–52, 56–57

Von Haller, Albrecht, 29–30

Wade, Nicholas, 125, 126, 207

Watkins, Elizabeth, 158

Watson, James, 101, 123, 131, 137, 174, 187, 190–193; as Associate Director for Human Genome Research, 192; as coordinator of Human Genome Project at NIH, 190, 191; as Director of the National

Watson, James *(continued)*
    Center for Human Genome Research, 192; ELSI provision and, 199–200; leaves NIH, 200, 201
Webber, Herbert John, 120
Weinberg, Robert A., 227
Weiner, Charles, 138–139
Weismann, August, 73–74, 87, 246
Weiss, Rick, 217, 290
Weissman, Gerald, 205
Weissman, Irving, 270, 271
Wellcome Trust, 171
Wells, Deane, 144, 151
West, Michael, 269
Wexler, Nancy, 196, 197–198
When life begins (debate), 2, 4–5, 9, 30–31, 54, 55–56, 57, 59, 82, 138, 139, 211, 240, 301; at capacity to feel, 262; cloning issues and, 235–236; at conception/fertilization, 3, 7–9, 25, 42, 235, 257; court rulings, 162, 163; developmental stages (gradualist view), 8, 9, 10–11, 13–16, 23, 24, 25, 80, 246; at development of vital organs, 11; ensoulment theory (and quickening), 7, 14–15, 17, 18–19, 22, 25, 31, 235; at fortieth day, 17, 18–19, 31, 56, 65–67, 163, 262; at gastrulation, 261; historical overview, 7–8, 9, 11, 12; 16–19, 31; at mixing of male and female fluids, 7, 21, 23, 25, 42, 43; potential person versus actual person, 11, 15, 41, 162–163; at presence of fetal heartbeat, 9, 29–30, 262
Whitman, Charles Otis, 57, 99
Wick, Tessa, 6, 7
William, William Carlos, 109
Wilmut, Ian, 215–217, 219–223, 236, 242
Wilson, Edmund Beecher, 46–49, 88, 99, 253
Wolff, Caspar Friedrich, 29
Wolpert, Lewis, 261
Women's movement, 140, 161, 164
World Health Organization, 161
Wyngaarden, James, 186, 190

Yale University, 159–160

Zucker, Katie, 6–7